Analysis of the Magnetic Field Observations of the Comet Lander Philae

Analysis of the Magnetic Field Observations of the Comet Lander Philae

Von der Fakultät für Elektrotechnik, Informationstechnik, Physik
der Technischen Universität Carolo-Wilhelmina zu Braunschweig

zur Erlangung des Grades eines Doktors
der Naturwissenschaften (Dr. rer. nat.)

genehmigte Dissertation

von Philip Thomas Heinisch

aus Uelzen

eingereicht am: 19.02.2019

Disputation am: 29.05.2019

1. Referent: Prof. Dr. Karl-Heinz Glaßmeier
2. Referent: Prof. Dr. Jürgen Blum

Druckjahr: 2019

Bibliografische Information der Deutschen Nationalbibliothek
Die Deutsche Nationalbibliothek verzeichnet diese Publikation in der Deutschen
Nationalbibliographie; detaillierte bibliographische Daten sind im Internet über
http://dnb.d-nb.de abrufbar.
1. Aufl. - Göttingen: Cuvillier, 2019
Zugl.: (TU) Braunschweig, Univ., Diss., 2019

Dissertation an der Technischen Universität Braunschweig,
Fakultät für Elektrotechnik, Informationstechnik, Physik

© Cover-Foto: ESA/ATG medialab; Comet image: ESA/Rosetta/Navcam

© CUVILLIER VERLAG, Göttingen 2019
Nonnenstieg 8, 37075 Göttingen
Telefon: 0551-54724-0
Telefax: 0551-54724-21
www.cuvillier.de

ISBN 978-3-7369-7040-3
eISBN 978-3-7369-6040-4

Veröffentlichungen

Teilergebnisse aus dieser Arbeit wurden mit Genehmigung der Fakultät für Elektrotechnik, Informationstechnik, Physik, vertreten durch den Mentor der Arbeit, in folgenden Beiträgen vorab veröffentlicht.

- Jurado, E., Martin, T., Canalias, E., Blazquez, A., Garmier, R., Ceolin, T., Gaudon, P., Delmas, C., Biele, J., Ulamec, S., Remetean, E., Torres, A., Laurent-Varin, J., Dolives, B., Herique, A., Rogez, Y., Kofman, W., Jorda, L., Zakharov, V., Crifo, J.-F., Rodionov, A., **Heinisch, P.**, Vincent, J.-B., Rosetta lander Philae: Flight dynamics analyses for landing site selection and post-landing operations, Acta Astronautica, 2016, 125, 65 – 79, ISSN 0094-5765

- Richter, I., Auster, H.-U., Berghofer, G., Carr, C., Cupido, E., Fornaçon, K.-H., Goetz, C., **Heinisch, P.**, Koenders, C., Stoll, B., Tsurutani, B. T., Vallat, C., Volwerk, M., Glassmeier, K.-H., Two-point observations of low-frequency waves at 67P/Churyumov-Gerasimenko during the descent of Philae: comparison of RPC-MAG and ROMAP, Annales Geophysicae, 2016, 34, 609 – 622

- **Heinisch, P.**, Auster, H.-U., Plettemeier, D., Kofman, W., Herique, A., Statz, C., Hahnel, R., Rogez, Y., Richter, I., Hilchenbach, M., Jurado, E., Garmier, R., Martin, T., Finke, F., Güttler, C., Sierks, H., Glassmeier, K.-H., Reconstruction of the flight and attitude of Rosetta's lander Philae, Acta Astronautica, 2017, 140, 509 – 516, ISSN 0094-5765

- **Heinisch, P.**, Auster, H.-U., Richter, I., Haerendel, G., Apathy, I., Fornacon, K.-H., Cupido, E., Glassmeier, K.-H., Joint two-point observations of LF-waves at 67P/Churyumov-Gerasimenko, Monthly Notices of the Royal Astronomical Society, 2017, 469, 68 -- 72

- **Heinisch, P.**, Auster, H.-U., Gundlach, B., Blum, J., Güttler, C., Tubiana, C., Sierks, H., Hilchenbach, M., Biele, J., Richter, I., Glassmeier, K. H., Compressive strength of comet 67P/Churyumov-Gerasimenko derived from Philae surface contacts, Astronomy & Astrophysics, 2018, im Druck

- **Heinisch, P.**, Auster, H.-U., Richter, I., Glassmeier, K.-H., Revisiting the magnetization of comet 67P/Churyumov-Gerasimenko, Astronomy & Astrophysics, 2018, im Druck

- Biersteker, J. B., Weiss, B. P., **Heinisch, P.**, Herčik, D., Glassmeier, K.-H., Auster, H.-U., Implications of Philae magnetometry measurements at comet 67P/Churyumov-Gerasimenko for the outer solar system nebular field, The Astrophysical Journal, 2019, 875, 1

Contents

Contents

Abstract

On November 12, 2014 the Philae spaceprobe separated from the Rosetta host spacecraft and touched down on comet 67P/Churyumov-Gerasimenko. The two spacecrafts arrived at the comet in August 2014 as part of the ESA Rosetta mission. After the initial touch-down of Philae the anchoring systems failed and the lander bounced of the surface and came to a final stop below a cliff after three additional surface contacts. Because of unfavorable lighting conditions not enough power was available to immediately recharge the batteries as intended. Radio contact with the lander was lost on November 15, 2014 at 00:36 UTC after the batteries were depleted. The Rosetta orbiter remained in orbit for two years to study the evolution of the comet and the plasma environment through perihelion in August 2015. After three mission extensions the orbiter was guided down to the surface of the comet in September 2016, officially ending the Rosetta mission.

This thesis focuses on the operations of the Philae lander, especially the measurements obtained by the lander magnetometer ROMAP. During descent and after touchdown both the orbiter magnetometer RPC-MAG and ROMAP were operating simultaneously, which allowed for in-situ magnetic two-point observations. These observations were used to determine the characteristics of the low-frequency "singing comet" waves present in the nucleus surface region including frequency, propagation direction and velocity. The results showed, that the waves propagate predominantly from the direction of the nucleus towards the Sun with a mean phase velocity of ∼ 5.3 km/s, a wavelength of ∼ 660 km and a frequency range between 5 mHz and 50 mHz.

The concurrent measurements were also used to determine the dynamics and attitude of Philae during descent and rebound. With these results an updated upper limit of 0.9 nT for any contribution from surface magnetization to the observed magnetic field was derived. Hence, the previous estimation of the surface magnetization was revisited. Based on the new knowledge about the lander rebound it was also possible to improve the achieved spatial resolution for the derived magnetization down to the size of individual aggregates in the range of ∼ 5 cm. For homogeneously magnetized pebbles in this size range the limit of 0.9 nT translates to an upper limit of ∼ $5 \cdot 10^{-5}$ Am^2/kg for the specific magnetic moment. Based on available magnetization models and depending on the exact history of formation, the newly achieved spatial resolution allowed to constrain the magnitude of the background magnetic field in the solar nebular during the formation of the comet to below 4 μT.

By combining the descent and rebound reconstruction with images from the orbiter camera system possible rebound trajectories of Philae were determined. Based on this flight reconstruction the energy balance of Philae's flight was derived. This analysis showed that a pressure of ∼ 100 Pa is enough to compress the surface material up to a depth of ∼ 20 cm. Considering all errors the derived compressive strength shows little dependence on location with an overall upper limit for the surface compressive strength of ∼ 800 Pa.

Kurzfassung

Am 12. November 2014 wurde die Landeeinheit Philae von der Rosetta Sonde abgesetzt und landete auf dem Kometen 67P/Churyumov-Gerasimenko. Beide Raumfahrzeuge erreichten den Kometen im August 2014 als Teil der ESA Rosetta Mission. Nach dem ersten Aufsetzen versagte das Ankersystem, der Lander prallte ab und kam nach drei weiteren Oberflächenkontakten unter einem Kliff zum Liegen. Durch die schlechten Beleuchtungsverhältnisse war nicht ausreichend Solarenergie verfügbar um die Akkus wie geplant aufzuladen. Daher brach der Funkkontakt mit dem Lander am 15. November um 00:36 UTC ab. Rosetta blieb für weitere zwei Jahre im Orbit um die Entwicklung des Kometen und der Plasmaumgebung während und nach dem Perihel-Durchgang zu untersuchen. Nach drei Verlängerungen endete die Rosetta Mission offiziell im September 2016 mit dem Aufsetzen des Orbiters auf dem Kometen.

Diese Arbeit behandelt den Philae Lander, speziell die Messungen mit dem ROMAP Magnetometer. Während des Abstiegs und nach der Landung waren sowohl das Orbiter Magnetometer RPC-MAG als auch das ROMAP Magnetometer gleichzeitig aktiv und ermöglichten simultane Zweipunkt - Messungen. Diese erlaubten es, die Frequenz, Ausbreitungsrichtung und Geschwindigkeit der oberflächennahen niederfrequenten "singing comet" Plasmawellen zu analysieren. Die Ergebnisse zeigten, dass sich diese Wellen primär aus der Richtung des Nucleus zur Sonne mit einer mittleren Phasengeschwindigkeit von ~ 5.3 km/s und einer Wellenlänge von ~ 660 km ausbreiten. Der typische Frequenzbereich ist 5 mHz bis 50 mHz.

Die gleichzeitigen Messungen wurden darüber hinaus verwendet, um die Dynamik und Orientierung des Philae Landers während des Abstiegs und des weiteren Flugs nach dem Abprallen zu bestimmen. Mit diesen Ergebnissen war es möglich, einen neuen Grenzwert von 0.9 nT für den Beitrag einer möglichen Oberflächenmagnetisierung zum gemessenen Magnetfeld abzuleiten. Daher wurde im Rahmen dieser Arbeit die bisherige Abschätzung der Magnetisierung aktualisiert. Basierend auf dem rekonstruierten Flug nach dem Abprallen war es auch möglich, die bisherige räumliche Auflösung der bestimmten Magnetisierung bis auf die Größenordnung einzelner Aggregate im Bereich von ~ 5 cm zu verbessern. Für solche Partikel bedeutet die Grenze von 0.9 nT ein maximales spezifisches magnetisches Moment von ~ $5 \cdot 10^{-5}$ Am^2/kg. Basierend auf verfügbaren Modellen zur Magnetisierung und abhängig von der Entstehungsgeschichte ermöglicht die feinere räumliche Auflösung, die Stärke des Hintergrundmagnetfelds im solaren Nebel während der Entstehung des Kometen auf unter 4 μT einzugrenzen.

Durch eine Kombination der Flugrekonstruktion mit Bildern des Kamerasystems an Bord des Orbiters konnten mögliche Flugbahnen nach dem Abprallen bestimmt werden. Dies ermöglichte es, die Energiebilanz des Flugs zu bestimmen. Diese Analyse zeigte, dass bereits ein Druck von ~ 100 Pa ausreichend ist, um das Oberflächenmaterial des Kometen bis zu einer Tiefe von ~ 20 cm zu komprimieren. Unter Berücksichtigung der Fehler zeigte die Druckfestigkeit nur eine geringe Abhängigkeit vom Ort und es ergab sich eine maximale obere Schranke von ~ 800 Pa.

1 Introduction

Comets have always fascinated mankind and sparked scientific interest. Because their relatively fast movement across the sky is apparent even with basic instruments, they were one of the first objects used to experimentally study the motion of celestial bodies. Most notably comet C/1680 V1 Kirch passing the earth at 0.42 AU in 1680 spurred interest in finding a mathematical base for Kepler's laws of planetary motion (Lancaster-Brown 1985). Isaac Newton succeeded in calculating the orbit of comet Kirch and later published an extended version of his mathematical framework in 1687 (Newton 1687), revolutionizing modern physics. Edmond Halley, who arranged for the publication of Newtons work, later used the results himself to calculate the orbit of comet 1P/Halley (Halley 1705). This comet became the target of five different satellites (the so called Halley Armada) during its apparition 281 years later in 1986. These space probes were launched in a coordinated effort by the Soviets, ESA and Japan to perform flybys and study 1P/Halley in situ (e.g. Grewing et al. 1988). Additionally NASA intended to support observations with two space shuttle missions. Being the first time close-up images of a small solar system object were taken by space probes, it marked a great technological and scientific achievement. The success of the Halley Armada demonstrated the capabilities of unmanned space probes and the scientific importance of in situ observations. Shortly after ESA and NASA started planning possible follow-up missions. After several delays and unsuccessful attempts to secure the necessary funding, ESA finally approved the Rosetta mission in 1993 (Glassmeier et al. 2007a). Consisting of the Rosetta orbiter and the Philae lander and being supported by NASA, the space probes were launched in March 2004 aboard an Ariane 5 rocket. After arriving at the target comet 67P/Churyumov–Gerasimenko (in the following shortened to 67P) in 2014 and delivering the lander to the surface, the mission was successfully concluded in 2016 after several extensions. The Rosetta mission achieved several scientific and technological firsts. Most notably the in situ study of a comet around perihelion and the controlled landing of a space probe on a small solar system object.

In this thesis, the measurements returned by Philae after separation from the Rosetta orbiter on November 12, 2014 until radio contact was lost on November 15 are analyzed and put in context. The focus lies on the magnetic field observations provided by the ROMAP magnetometer (Auster et al. 2007) onboard Philae and the concurrent measurements of the orbiter magnetometer RPC-MAG (Glassmeier et al. 2007b). Based on these observations, the magnetic field in the plasma environment around the nucleus is characterized, and upper limits for the surface magnetization and compressive strength are derived. Additionally the magnetic measurements are used to reconstruct the lander dynamics and attitude and provide status information about the internal lander systems.

The second chapter provides an overview about the Rosetta mission, the Philae lander and comet 67P. Because of the importance for this study, the lander instrument ROMAP is described in detail. Chapter three gives an introduction to the cometary plasma environment and the typical accompanying magnetic field structures. One of the most striking magnetic features during the joint Rosetta and Philae mission in fall 2014 were low frequency magnetic waves (Richter et al. 2015, 2016). These newly detected waves were analyzed using concurrent two-point observations from Philae on the surface and Rosetta in orbit above the comet. At the end of the chapter it is explained, how magnetic field measurements can also be used to determine the attitude of Philae and derive the lander dynamics.

In the following, the magnetic field observations already analyzed in the previous chapter are combined with images and status information from the solar arrays and radio communication equipment to determine in detail what happened to Philae during its flight above the surface of 67P. In addition to a description of the descent dynamics and the attitude during rebound, the approximate coordinates for the surface contacts are estimated. It will be shown, that Philae did not change attitude between the end of the initial measurements in 2014 and the discovery of the lander on Rosetta images in 2016.

The flight reconstruction in conjunction with the magnetic field analysis performed in chapter three is used in chapter five to derive an upper limit for the surface magnetization of the comet. While the magnetization has been investigated before, this work makes use of the comprehensive understanding of the circumstances of Philae's descent and landing, gained after the end of the mission, to revisit the magnetic properties.

In chapter six the previous results were combined to approximate the descent and rebound trajectory of the Philae lander and use this information to derive the compressive strength of the surface material from the different surface contacts and scratches created during the final touchdown. One of the primary objectives of the lander mission was to measure the surface properties (such as the compressive strength) of 67P in situ to determine what comets are made of and where they formed. While the growth of kilometer-sized planetesimals to larger planet-sized objects in the early Solar System is well understood, it is still under debate how smaller objects formed out of sub-decimeter-sized dust aggregates (see Blum (2018) for a recent review). Because of their primitive nature, comets are the best candidates for planetesimals and possibly the sole small survivors of the planet-formation era (Davidsson et al. 2016). The mechanical strength of the cometary material strongly depends on the formation history and evolution of the nucleus (Blum et al. 2014). Thus, measurements of mechanical properties combined with already available laboratory results can be used to study the origin and evolution of comets and can thereby provide the missing link between protoplanetary dust and planets.

2   The Rosetta Mission

Rosetta was an ESA cornerstone mission to study the periodic comet 67P. It was launched on March 2, 2004 as part of ESA's Horizon 2000 program with support from NASA. The initial target for the Rosetta mission was comet 46P/Wirtanen, but due to uncertainties concerning the reliability of the Ariane 5 system, the launch was postponed and 67P selected as the new target (Ulamec et al. 2006). One of the major objectives of the mission was to deploy the Philae lander to allow scientific measurements on the surface. While the Rosetta spacecraft was managed by ESA, Philae was developed and build by a consortium headed by the German space agency DLR and the french CNES. In contrast to previous missions, Rosetta should not only collect data during short flybys, but be the first space probe to rendezvous with a comet and closely follow it for more than a year. After being the first man made craft to land on a comet, Philae was intended to take different measurements on the surface for several days.

2.1   Rosetta Overview

Figure 2.1 gives an overview of the trajectories of comet 67P and Rosetta during its mission. The spacecraft reached 67P shortly after it passed aphelion in August 2014 at a heliocentric distance of more than 3.6 AU, after a 31 month deep space hibernation phase. To match the speed of the comet Rosetta performed several gravity assist manoeuvres. Flying this trajectory required additional time, stretching the journey to about ten years. Using chemical boosters to accelerate instead, would have required prohibitively large fuel supplies. Electrical propulsion systems, while being capable of providing long term thrust with little fuel, are unsuitable for missions like Rosetta due to the limited electrical power budget.

In the outer solar systems at distances of more than 5.3 AU the generated solar power was only sufficient to supply basic survival heaters to keep the spacecraft from freezing. Hence, it was necessary to put Rosetta into a special hibernation mode during the cruise phase from 2011 until 2014 to conserve power. Before entering hibernation, Rosetta performed flybys of asteroids 2867 Steins and 21 Lutetia (Barucci et al. 2007). These flybys returned initial scientific results even during the cruise phase (Auster et al. 2010, Keller et al. 2010, Schulz 2010, Schulz et al. 2012, Richter et al. 2012) an allowed in-flight testing and calibration of both lander and orbiter instruments.

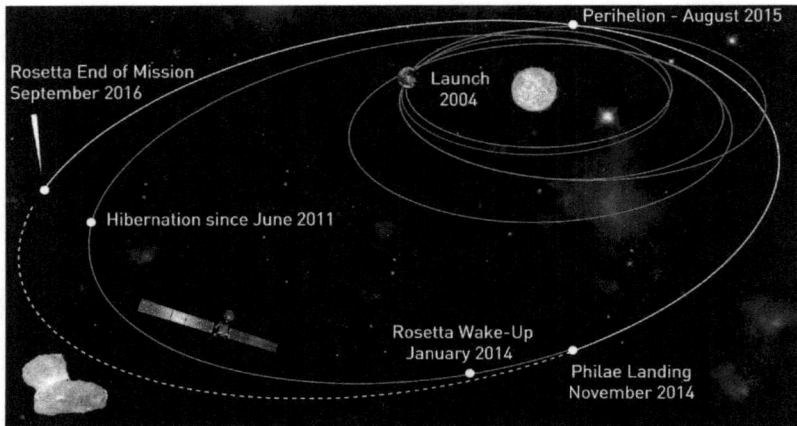

Figure 2.1: Illustration of the orbits of comet 67P (white), Rosetta (red) and the earth (blue) during the progression of the Rosetta mission (parts of this illustration were released by ESA under the Creative Commons Attribution Share Alike 3.0 license).

In November 2014, Rosetta deployed the Philae lander and the orbiter operations were primarily aimed at supporting the lander mission for several weeks (Ashman et al. 2016). Before deployment, concurrent observations from different Rosetta instruments were used to map the comet in preparation of the landing site selection process (Jurado et al. 2016). Afterwards, the antennas onboard Rosetta had to be pointed at the landing site of Philae to communicate with the lander, leading to significant pointing constraints. As the exact position of Philae after landing was unknown, the area around the suspected position was photographed from orbit in an attempt to locate the lander. During this time Rosetta remained in a terminator orbit close to the nucleus, at a mean distance of 38 km. A terminator orbit is defined as a trajectory above the division between the illuminated and dark hemispheres. In early 2015, Rosetta started to perform close flybys at distances going down to 6 km. Closer to perihelion in the spring of 2015, stronger insolation increased the activity of 67P causing more dust to be ejected from the surface. These particles interfered with the star tracker cameras (Buemi et al. 2000) used by Rosetta to automatically determine its position and attitude. The effect of the dust on the tracking cameras was worse than expected and the cometocentric distance had to be increased significantly to ensure safe operations.

To study different regions of the plasma environment (Volwerk et al. 2018), Rosetta performed excursions with significantly increased cometocentric distance (> 1500 km). After several extensions, the Rosetta mission ended on September 30, 2016 with the controlled impact of the Rosetta probe on the comet. During its two years around 67P, Rosetta collected enough scientific measurements that a comprehensive analysis will take years (Barthelemy et al. 2018).

2.2 Overview of the Philae Landing

On November 12, 2014 at 08:35 UTC Philae was detached from Rosetta and touched down at 15:35 UTC, after a controlled descent. After separation, Philae descended on a ballistic trajectory towards the surface, landing 7 h later. This first part of the Philae operations, known as separation, descent and landing (SDL), was executed as planned and Philae operated flawlessly. During the descent in addition to a farewell and several descent images (Mottola et al. 2015), only ranging and magnetic field measurements were performed to conserve telemetry bandwidth and power. Accelerometers in the feet were automatically triggered during the initial surface contact, to measure the shock of the touchdown. After touchdown, Philae was intended to automatically anchor itself to the surface using harpoons and ice screws, while a cold gas thruster should have provided hold-down thrust (see section 2.5.1 for a description of the lander subsystems). All these systems failed, causing Philae to rebound (Ulamec et al. 2016). Nevertheless the touch-down signal automatically started the initial sequences of preprogrammed measurements called the first science sequence (FSS). These observations were intended to be performed after Philae was stationary and secured to the surface. As the lander was instead in-flight, the scientific return of these initial FSS measurements was limited.

After three bounces the still operational lander came to a final stop approximately 1.3 km away from the intended landing site (for a detailed description of the trajectory see section 4). The regular sequence of measurements was remotely terminated and instead replaced by several so called "mechanical safe blocks". During these sequences only instruments without mechanical components (i.e. no drilling or hammering) were operated. In addition to camera images and thermal mapping, ranging and magnetic field measurements were performed. While radio contact was still possible, Philae was below a cliff-like structure, shadowing the solar arrays in an already sparsely lit area. Therefore the solar panels could not provide enough power to simultaneously heat and recharge the batteries (Ulamec and Taylor 2016). Because of the chemical characteristics of the lithium based batteries, the internal temperature had to be $> 10°$ C to allow for safe recharging. This should have allowed for several days or even month of lander operations (Bibring et al. 2007b), before the rising internal temperature would have caused the lander to fail, as it got closer to the sun.

Initially it was unknown how stable Philae was without any anchoring, hence the intention of the safe blocks was to prevent mechanical systems from accidentally moving, tilting or even tipping the lander. These unfortunate circumstances only allowed for minimal scientific surface operations and prevented the use of some of the main instruments completely. It was deemed too dangerous to use the drill stack to sample the surface. Therefore the onboard mass spectrometers were not able to analyze the composition of the cometary material and were only operated in a passive sniffing mode (Goesmann et al. 2015). After both the primary and redundant radio transceivers onboard Philae started to malfunction, it became clear, that the planned long term science operation (LTS) was most likely not feasible. To measure at least some mechanical properties of the surface (mainly compressive strength and thermal conduction), the MUPUS hammer system (see section 2.5.1) was deployed shortly before the battery was depleted (Spohn et al. 2015).

Because of power and up-link data rate limitations, it was not possible to use the camera system to track the deployment of the hammer. Afterwards, the lander body was rotated around the landing gear to improve the illumination of the solar panels. Contact was lost on November 15, at 00:36 UTC when the battery drop-out voltage was reached.

Afterwards the Rosetta orbiter remained close the nucleus for several days, trying to reestablish radio contact and identify Philae on orbiter camera images to confirm the position. These initial attempts failed and due to increased activity of the comet, Rosetta was moved to a more distant orbit for safety reasons, making further radio contact nearly impossible. On June 13, 2015 a short signal from Philae was received by the Rosetta orbiter, indicating that Philae was still healthy and now, closer to the Sun, enough solar power was available for scientific operations. To increase the likelihood of further radio communications, the orbit and pointing of Rosetta was adapted based on the estimated location and attitude of Philae (Heinisch et al. 2016). After several successful contacts, attempts to activate scientific instruments failed and after July 9, no further communication was possible. On July 27, 2016 the Philae radio transceivers onboard Rosetta were switched off, officially ending Philae operations.

2.3   Comet 67P/Churyumov–Gerasimenko

67P is a short-period Jupiter family comet, discovered by Klim Churyumov and Svetlana Gerasimenko in 1969. The discovery itself and the orbital parameters are described in detail by Krolikowska (2003) and Lamy et al. (2007). 67P has a heliocentric perihelion distance of approximately 1.3 AU and an aphelion distance of 5.7 AU with an orbital period of 6.55 years. It rotates around an axis of largest moment of inertia with a period of 12.4 h (Mottola et al. 2014). While initial remote observations of 67P by Lamy et al. (2006), using the Hubble space telescope, suggested an elongated shape, similar to a potato, images taken by the Rosetta spacecraft revealed a much more intricate bi-lobed structure. Fig. 2.2 shows an image of the comet, taken by Rosetta at a distance of 154 km in July 2015 as an example. The smaller upper lobe is approximately 2.6 km by 2.3 km and has a height of 1.8 km, while the bigger bottom lobe is 4.1 km by 3.3 km and 1.8 km high (Mottola et al. 2014). This unexpectedly irregular shape made it much more difficult than initially expected to find a suitable landing site for Philae during the landing site selection process as explained by Ulamec et al. (2015) and Jurado et al. (2016). This was not only caused by the complex gravitational field of such an object, but also due to significant self shadowing. Entire regions of the comet were inaccessible to Philae as high cliff-like structures blocked possible trajectories. Because of the shape of 67P and orientation of the rotation axis, several different regions experienced polar day and night during the progression along the orbit of the comet. This led to inhomogeneous erosion of these areas and further complicated the selection of a landing site. Starting from an initial set of ten possible landing sites, the search was narrowed down to five sights with the best possible scientific return while at the same time providing the highest likelihood of longer-term lander survival. These possible areas are illustrated in Fig. 2.3. During the landing site selection process in fall 2014, site "J" was finally chosen as primary landing site for Philae.

Figure 2.2: Image of comet 67P/Churyumov–Gerasimenko taken by Rosetta on July 7, 2015 from a distance of 154 km (ESA/Rosetta/NAVCAM, this image was published under the Creative Commons Attribution Share Alike 3.0 license).

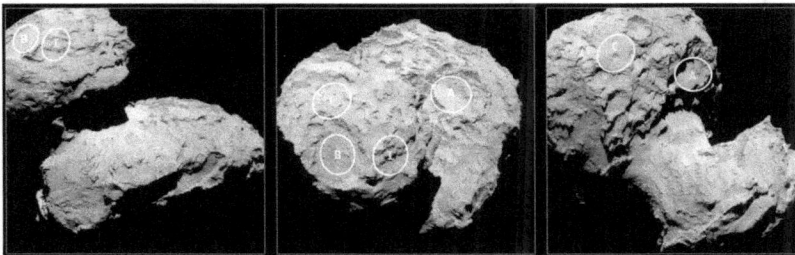

Figure 2.3: Five possible landing sites selected during the landing site selection process in 2014, site "J" was selected as primarily landing site for Philae (released for publication by ESA/Rosetta/MPS for OSIRIS Team MPS/UPD/LAM/IAA/SSO/INTA/UPM/DASP/IDA).

Figure 2.4: Rendering of the Rosetta spacecraft showing the (primarily) Sun-facing site. Image provided by ATG medialab for ESA (this image was published under the Creative Commons Attribution Share Alike 2.0 license).

Perturbations of Rosetta's orbital velocity allowed an accurate determination of the mass and bulk density. Using this approach Pätzold et al. (2016) derived a mass of $9.98 \cdot 10^{12}$ kg and an average bulk density of 533 kg/m^3. A total volume of 18.7 km^3 was estimated from camera images (Jorda et al. 2016), while an overall porosity of 75% to 85% was determined from radar observations (Kofman et al. 2015). By combining these results with observations from several other Rosetta instruments Davidsson et al. (2016) and Blum et al. (2017) independently concluded, that 67P most likely formed as a primordial rubble pile. This means that the comet was created by agglomeration of smaller pebbles and not as a collisional fragment of a larger parent body as suggested for example by Morbidelli and Rickman (2015). The nature of 67P makes it an ideal target to study the evolution of remnants of the early solar system.

2.4   The Rosetta Spacecraft

The Rosetta spacecraft was purpose-build for the very specific demands of a cometary mission. Reliable operation for at least ten years and thermal management and power systems capable of supporting spacecraft survival in deep space at more than 5 AU were the main requirements. To accomplish this without the use of nuclear power systems, Rosetta was equipped with 64 m^2 of solar cells, which was the largest solar array ever flown (D'Accolti et al. 2002) up to this point. The main spacecraft bus was build by Astrium (now Airbus) and measured 2.8 m by 2.1 m with a height of 2 m and a total mass of approximately 2900 kg. Fig 2.4 shows a rendering of the spacecraft, illustrating the dimensions and the prominent solar arrays.

After launch on March 2, 2004 Rosetta performed several gravity assist maneuvers at Earth and Mars and flew past Asteroids 2867 Steins (Schulz 2010) and 21 Lutetia (Schulz et al. 2012) in September 2008 and and July 2010 respectively. In June 2011 Rosetta and Philae entered a deep space hibernation mode (Ulamec et al. 2012, Ferri et al. 2012), disabling nearly every system except an automated wake up timer and survival heaters to keep the spacecrafts from freezing. After 31 month of low power sleep mode, Rosetta woke up successfully, as expected, on January 20, 2014 at a heliocentric distance of 4.5 AU.

To allow for scientific observations, Rosetta was equipped with eleven scientific instruments. In addition to camera systems, spectrometers (Coradini et al. 2007, Stern et al. 2007, Balsiger et al. 2007, Gulkis et al. 2007), dust analyzers (Colangeli et al. 2007, Kissel et al. 2007, Riedler et al. 2007) and radio science instrumentation (Pätzold et al. 2007) Rosetta was also equipped with a suite of plasma instruments. In addition to the scientific instruments, several different sensors primarily for operational purposes like accelerometers, navigation cameras and radiation monitors were included. While the results were mainly used for planning and navigation, they were also used to complement the scientific measurements. To interpret the magnetic field measurements and analyze Philae's flight, observations from different Rosetta instruments were combined. For this purpose, the most important instruments beside the magnetometers were the camera system and radar instrument. Hence, a short description of these sensors is given in the following section.

In the context of this study, the most important instruments of the Rosetta orbiter were the two magnetometers as part of the Rosetta Plasma Consortium (RPC) described in detail by Carr et al. (2007). In addition to the magnetometer RPC-MAG, RPC was comprised of four other instruments and a central power handling system called Plasma Interface Unit (PIU). The RPC-MAG magnetometer (Glassmeier et al. 2007b) consists of two tri-axial fluxgate magnetometers, an outboard sensor (RPC-OB) with a maximum sampling rate of 20 Hz and an inboard sensor (RPC-IB) with a maximum sampling rate of 1 Hz. To reduce spacecraft interference, the magnetometers were mounted 15 cm apart on a boom at a distance of 1.5 m from the main Rosetta body. Nevertheless a magnetic signature from the spacecrafts internal flywheels primarily at typical frequencies between 1 Hz and 10 Hz were present in the field measurements. As the interference signal was well characterized, it was possible to remove it from most of the observations used in the following. Especially due to constantly changing temperatures, offset calibration of the two sensors was challenging. During the Philae landing and surface operation, cross-calibration with the magnetic observations by the lander magnetometer was used to dynamically determine the RPC-MAG offsets (Richter et al. 2016).

Part of the RPC suite of instruments were also the Ion Composition Analyser (RPC-ICA), Ion and Electron Sensor (RPC-IES), Langmuir Probe (RPC-LAP) and the Mutual Impedance Probe (RPC-MIP). RPC-IES (Burch et al. 2007) and RPC-ICA (Nilsson et al. 2007) were designed to detect electrons and ions and determine the three dimensional spatial distribution. RPC-LAP (Eriksson et al. 2007) and RPC-MIP (Trotignon et al. 2007) were used to measure the plasma density, electron temperature, spacecraft potential and the plasma flow velocity. The Comet Pressure sensors (COPS) which were part of the Rosetta Spectrometer for Ion and Neutral Analysis (ROSINA - Balsiger et al. 2007) were also used together with the RPC sensor suite to provide information about the neutral gas pressure.

13

The Rosetta Optical, Spectroscopic and Infrared Remote Imaging System (OSIRIS) as described by Keller et al. (2007) was the main scientific camera system of the Rosetta orbiter. Two independent cameras, a near-angle camera (NAC) and a wide angle camera (WAC) were used. While the CCD imaging sensors were only black and white, different calibrated color filters made it possible to generate pseudo-color images and perform optical spectroscopy (Keller et al. 2007). Together with the wide angle navigational camera system (NAVCAM) OSIRIS was responsible for all images of the comet.

To investigate the interior of comet 67P the Comet Nucleus Sounding Experiment by Radiowave Transmission (CONSERT) was integrated into both the orbiter and lander (Kofman et al. 2007). It was implemented as a two part instrument, with a transmitter on the orbiter sending a radio frequency pulse with a wavelength of approximately 3 m to the CONSERT component on the lander, which retransmitted the signal back to the orbiter. Information about the relative position of the Philae lander and also about the inner structure of the nucleus can be gained from analyzing the time delay between transmission and reception of the signal. This is possible due to the low frequency of the RF pulses, which were able to penetrate through the cometary material.

2.5   The Philae Lander

The Philae lander was the surface science package of the Rosetta mission and named after the Philae obelisk (found near Aswan in Egypt) following the Rosetta naming scheme. Philae in its entirety was considered as one of the Rosetta instruments and as such independently designed, build and operated by an international consortium headed by the German aerospace agency DLR and the french space agency CNES (Bibring et al. 2007b). Philae had an overall mass of 97.63 kg, including 10 kg for the landing gear assembly and 27 kg of scientific instruments (Biele et al. 2015). The main body had an approximate size of 1 m by 1 m and was 0.8 m high. During the cruise phase, Philae was connected to Rosetta via the mechanical support system (MSS) while power and data transfer was handled by the electrical support system (ESS). Almost the entire side and front and most of the top of Philae was covered by solar cells, while the rear balcony was reserved for instruments. A rendering of the Philae lander is shown in Fig. 2.5. For scientific operations, Philae was equipped with ten instruments, including cameras, a drill stack, different spectrometers, a magnetometer and a radio science package.

The initial target of Philae was comet 46P/Wirtanen, which is significantly smaller compared to 67P leading to a difference in mass of about two orders of magnitude (Lamy et al. 1998). While this change in the planned mission had little influence on the Rosetta spacecraft, it strongly impacted the lander and required modification of the landing gear as explained by Ulamec et al. (2006) to handle the higher loads expected while landing on the bigger comet 67P.

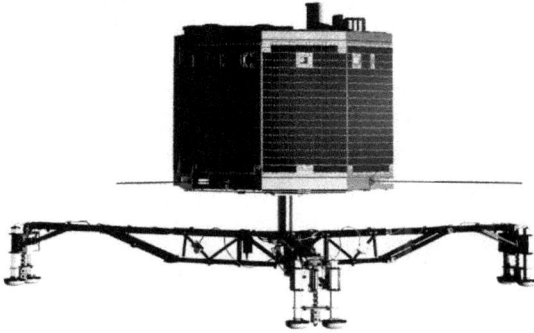

Figure 2.5: Rendering of the Philae lander viewed from the front, based on the mechanical model. Image provided by ATG medialab for ESA (this image was published under the Creative Commons Attribution Share Alike 2.0 license).

2.5.1 Lander Design

To have a mechanically stable platform Philae was equipped with a foldable landing gear with three legs. Two of these legs were 1.46 m in length while the front leg was only 1.42 m. This difference in length was necessary to fold the landing gear around the lander body during transit. Individual feet with two soles and an ice screw were attached at the end of the legs using pivoting joints to account for uneven terrain. The feet were in total ~ 28 cm high and 20 cm wide. Each of the soles were approximately 10 cm in diameter and 3.6 cm high. Fig. 2.6 shows an illustration of one of Philae's feet based on the 3D CAD model of the lander (provided by DLR as part of (Heinisch et al. 2016)). Additionally harpoons to anchor Philae to the surface (Thiel et al. 2003) were mounted on two of the legs. These harpoons were connected to the landing gear struts with retractable cables to provide tension after the harpoons were shot into the comet using pyrotechnic nitrocellulose gas generators. Accelerometers were fitted to the harpoons to record the deceleration while penetrating the surface to analyze the compressive strength.

One of each soles of the three feet were equipped with acoustic piezoelectric transducers as part of the SESAME suite of instruments (Seidensticker et al. 2007). These actuators (called SESAME-CASSE) were intended to introduce acoustic waves into the surface, which could be picked up by highly sensitive accelerometers mounted in the remaining soles to analyze seismic properties of the surface. These sensors were also used to record the acceleration of the lander during the initial surface contact (Arnold et al. 2014). Also part of this instrument package was a quadrupole permittivity probe (SESAME-PP) to measure the electric properties of the surface layers. A piezoelectric dust impact monitor (SESAME-DIM), mounted on the top plate of Philae was also part of the SESAME sensor suite.

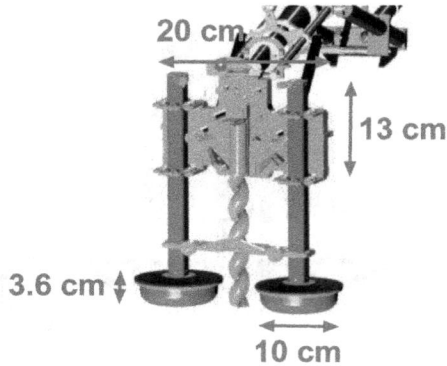

Figure 2.6: Illustration of one of Philae's feet based on a rendering of the 3D lander CAD model. The approximate dimensions of the soles and the upper feet structure are given in red.

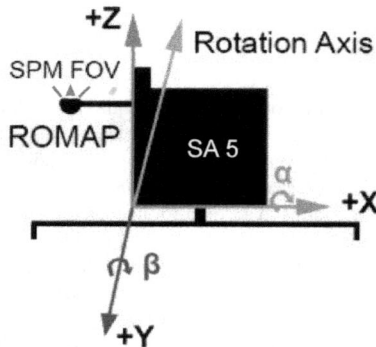

Figure 2.7: Illustration of the lander reference frame and rotation axis alignment angles α and β. The ROMAP SPM field-of-view is indicated by the funnel shaped lines and the arrow above the sensor (based on Heinisch et al. 2017a).

The baseplate of Philae to which the instruments and solar array hood was mounted, was linked to the landing gear with a universal cardan joint. This allowed rotation and tilting of the main body independently from the orientation of the landing gear. Thereby allowing attitude changes, even after the landing gear was anchored to the ground. The body fixed lander coordinate system was defined relative to the rear balcony edge of the baseplate, as illustrated in Fig. 2.7. The y-axis is parallel to the edge of the baseplate, the z-axis is perpendicular to the baseplate, pointing in the direction of the solar array hood and the positive x-axis is pointing toward the front solar panel.

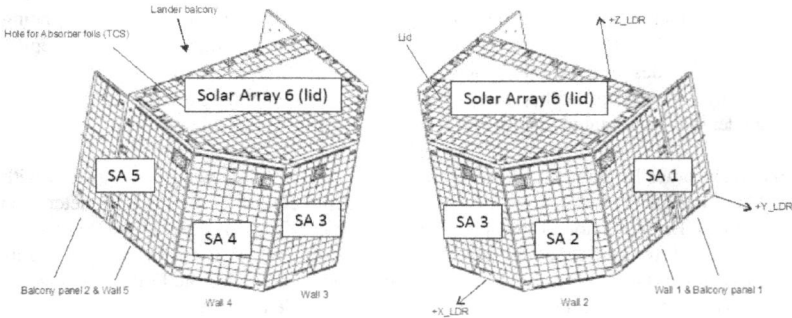

Figure 2.8: Illustration of the solar array hood of Philae showing the six solar panels and the lander reference coordinate system (adapted from DLR RB-TN-1035).

A damping mechanism was integrated into the tube connecting the landing gear joint to the base plate (see Fig. 2.5) as described by Witte et al. (2014). It consists of a spindle drive linked to a dampening motor, which was electrically short-circuited via a low-ohm resistor. Positional feedback for this system was provided by a potentiometer measuring the movement of the spindle. This setup transformed the downward motion of Philae during touchdown into rotation of the spindle, which turned the generator. The energy was than dissipated inside the connected resistor effectively turning the kinetic energy into heat without relaying on complex mechanical dampers. While this system is effective at damping the touchdown impact (Witte et al. 2016), uncertainties in the amount of kinetic energy dissipated as heat makes it more complicated to reconstruct mechanical properties based on the impact dynamics (Faber et al. 2015, Biele et al. 2015).

To provide hold-down thrust during touchdown and possibly increase the descent velocity, Philae was equipped with an active descent system (ADS) consisting of a nitrogen cold gas thruster build into the top plate. This thruster was programmed to activate automatically directly after touchdown was detected to counteract the harpoon recoil and prevent a rebound in case the anchoring with the harpoons and ice screws failed (Witte et al. 2016). Even though the active descent system could have provided additional acceleration during descent, the spindle drive of the MSS was enough to provide the planned 1 m/s relative velocity. To communicate with the orbiter, Philae was equipped with a redundant radio communication system, consisting of two panel antennas mounted on the top plate above the balcony and two redundant transceiver units. At nominal conditions, the system was designed to allow for a transmission range of at least 150 km (Dudal and Loisel 2016). For power generation Philae was outfitted with a solar array hood made up of six individual solar panels as can be seen in Fig. 2.5. Under cometary conditions these panels were able to provide a total electrical power of about 33 W. The individual panels were able to provide between 4.3 W and 6.7 W of power depending on their size (D'Accolti et al. 2002) and were labeled from solar array SA 1 to SA 6 as illustrated in Fig. 2.8. To save space and weight SA 1 and SA 5 shared the same maximum power point tracker and power converter, as they can not be illuminated at the same time due

17

to their location on opposite sites of Philae. Therefore only combined current and voltage measurements for housekeeping were available for SA 1 and SA 5 (therefore being appreciated as SA51), while each of the remaining panels had individual housekeeping channels. This design decision impacted the reconstruction of the Suns position based on panel power (described in chapter 4) as it made it impossible to distinguish which site of the lander was illuminated just based on SA 1 and SA 5.

Like many other landers (i.e. Mars, Gellert et al. 2006), Philae was also equipped with an Alpha Particle X-Ray Spectrometer (APXS - Klingelhöfer et al. 2007) to determine the elemental composition of the upper surface material. The sensor was mounted on two rods deployed from the baseplate of Philae. The instrument was to be lowered to the surface after touchdown, to accommodate for the unknown height of the Philae body above the surface, while also allowing for the lander to be folded up while in transit. Tests of the deployment mechanisms during commissioning of the spacecraft showed no movement of the rods. Failure of the deployment mechanism was also confirmed by magnetic field measurements, as no changes in the magnetic field either caused by the deployment motors or changes in the static bias field due to movement of the APXS instrument were detected. Subsequently the APSX instrument failed to deploy after touchdown and could not carry out any scientific measurements.

Philae was equipped with a total of eight optical cameras, one of them was the down looking ROLIS (Rosetta Lander Imaging System - Mottola et al. 2007) camera, while the rest of the cameras were part of the panoramic comet infrared and visible analyzer system (CIVA - Bibring et al. 2007a). Five of the panoramic CIVA-P cameras can be identified as rectangular recesses in the solar panels in Fig. 2.5 and 2.8, the remaining two were arranged as stereo cameras and mounted on the rear balcony. While the Rosetta orbiter was equipped with dedicated autonomous star tracker cameras (Buemi et al. 2000) for automatic attitude and position determination, the Philae lander had no such systems, due to constraints in weight and available power. For Philae no dedicated navigation instruments were deemed necessary, as the lander was to descent on a predetermined ballistic trajectory with line of sight to the orbiter. Therefore the intention was to use the orbiter cameras to determine the position of Philae after touching down on the comet.

To sample the cometary material, Philae was equipped with the sampler, drill and distribution system "SD2" (Finzi et al. 2007) consisting of a drill capable of extracting samples from a depth of up to 230 mm. These samples could than be stored in an internal carousel equipped with 26 ovens to heat the material and transfer the released gases to the two onboard spectrometers COSAC (Goesmann et al. 2015) and PTOLEMY (Wright et al. 2007). It was also possible to use two miniaturised microscopes (CIVA-M) from the CIVA instrument package to optically analyze the samples (Bibring et al. 2007a).

To derive the mechanical properties the multi purpose sensor package "MUPUS" (Spohn et al. 2007) consisted of a hammering mechanism, a penetrator and a deployment system mounted on the rear lander balcony. Fig. 2.9 shows an image of the stowed MUPUS hammering system mounted to the Philae ground reference model. The main hammer mechanism can be seen at the top center of the image connected to the carbon fiber pen-

Figure 2.9: Image of the stowed MUPUS (foreground) and ROMAP (bottom right above the baseplate) instruments mounted to the Philae ground reference model. The tip of the MUPUS penetrator is to the left of the ROMAP sensor head, just above the baseplate (adopted from Grygorczuk et al. 2011).

etrator rod, with the tip of the penetrator at the bottom directly above the baseplate. A down-looking thermal mapper based on four thermopile-type IR sensors (MUPUS-TM) was also mounted on the balcony, next to the CIVA-P cameras. The penetrator was to be deployed together with the hammering mechanism and placed on the surface. Afterwards the deployment mechanism was intended to retract to allow for later rotation of the lander body and provide mechanical decoupling. The deployment process should have been validated by images from the stereo CIVA-P cameras. The hammering system was than intended to drive the penetrator into the surface to determine the compressive strength, while the SESAME-CASSE sensors were to measure the corresponding seismic activity. Temperature sensors and heating elements were integrated inside the penetrator to measure the thermal properties of the cometary material.

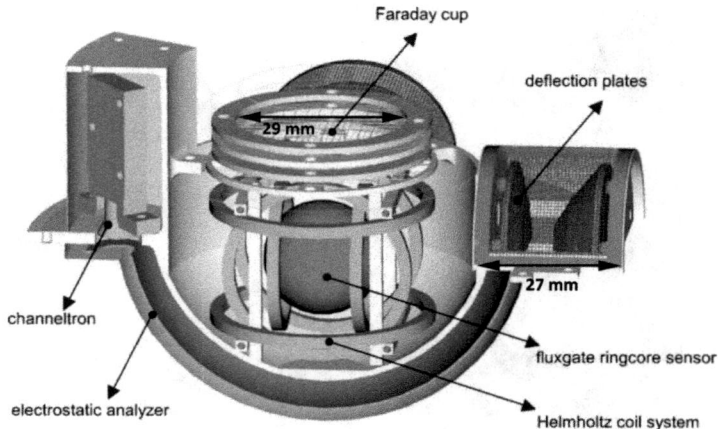

Figure 2.10: Schematic drawing of the ROMAP sensor, showing the main components of the fluxgate magnetometer and simple plasma monitor (based on Auster et al. 2015).

2.5.2 The ROMAP Instrument

The Rosetta Lander Magnetometer and Plasma Monitor (ROMAP) was a combined magnetometer and plasma monitor specifically designed for the Philae lander (Auster et al. 2007) to work in conjunction with the RPC sensor suite. A tri-axial fluxgate sensor with a maximum sampling rate of 32 Hz was used as magnetometer, while the simple plasma monitor (SPM) consisted of an hemispherical electrostatic analyzer with an energy range between 0 keV and 8 keV and a Faraday cup with an upper limit of 2 keV to study the plasma environment around the comet. Two different pressure sensors with a combined range of 10^{-8} hPa to 10 hPa were also part of the ROMAP package. Because of the unexpected circumstances of the landing and technical limitations, these sensors were not used during the surface operations.

A schematic drawing of the spherical ROMAP sensor head is displayed in Fig. 2.10 showing the Helmholtz coils (orange) around the Fluxgate magnetometer (red) in the middle below the retarding grid of the Faraday cup. The hemispherical electrostatic analyzer surrounds the Fluxgate sensor and the Faraday cup. Charged particles enter the hemispherical analyzer through deflection plates (blue plates in Fig. 2.10) on two sides of the sensor and depending on energy either get absorbed by the hemispherical analyzer channels or pass through and get detected by the channel electron multiplies (channeltrons) located opposite to the entrances. To protect the deflection plates, grids with a spacing of 2 mm are mounted on top of the entrance (red grids in Fig. 2.10). The aperture of the Faraday cup has a diameter of 29 mm and below it, a stack of four grid electrodes is located to select the energy level. The top retarding grid electrode has a spacing of 3

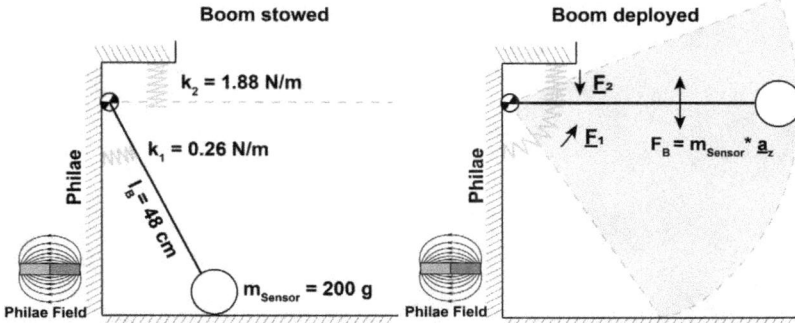

Figure 2.11: Schematic drawing of the ROMAP boom mounted to the Philae instrument balcony in the stowed (left) and deployed position (right).

mm, while the other electrodes have grid constants of 3 mm, 2 mm and 1 mm from top to bottom respectively.

To minimize the influence of lander bias fields on the magnetic field measurements, the ROMAP sensor head was mounted on a 48 cm long boom on the lander balcony (see Fig. 2.7). During flight the sensor and boom were stowed and launch locked to the lander balcony as can be seen in Fig. 2.9, showing a model of ROMAP mounted to the Philae ground reference model. After deployment the boom was to be kept in place by two springs, counterbalancing each other. The mechanics of the boom are illustrated in Fig. 2.11. Without a locking mechanism, the sensor was free to swing around the boom hinge as shown in the right hand panel of Fig. 2.11 after deployment. An assembly like this causes the magnetometer to swing from its nominal position as soon as the lander is accelerated perpendicular to the hinge axis. Movement of the boom causes a change in the observed lander bias field. This allows the boom to be used as a simple mechanical accelerometer along the lander z-axis (see Fig. 2.7). A qualitative assessment of the deflection angle and therefore the acceleration can be obtained by analyzing the change in the bias field observed by the magnetometer.

3 Magnetic Field around 67P/Churyumov–Gerasimenko

3.1 Cometary Plasma Environment

The plasma surrounding comets creates part of the most obvious cometary feature: the plasma tail. While the observation of the actual nucleus of a comet requires sophisticated telescopes, the cometary tail, comprised of the dust and plasma tail is detected easily by earth-based observations. The characteristic tail is the reason for the early discovery of comets compared to other small solar-system bodies, with first observations in China dating back to 1000 B.C. (Festou et al. 1993).

In contrast to the dust tail, the plasma tail always points radially outwards, away from the Sun, independent of the trajectory of the comet. This observable separation between the tail structures, exemplified in Fig. 3.1 by an image of comet 109P/Swift-Tuttle taken in 1892, lead Biermann (1951) to propose the concept of a solar wind as explanation. He established the solar wind as a stream of charged particles emanating from the Sun and propagating radially outward. The interaction between the solar wind and the material sublimating from the cometary nucleus is responsible for the creation of the plasma tail, as was later explained in detail by Alfvén (1957) and Biermann et al. (1967). They described this process in terms of mass, energy and momentum transfer between the solar wind plasma and the sublimated material inside the coma. The initially neutral molecules inside the coma get ionized primarily by solar UV radiation and particle-particle interactions (Heritier et al. 2018). Interaction between the newly created cometary plasma and the significantly faster solar wind leads to pick-up of the cometary ions, causing mass loading and a deceleration of the solar wind surrounding the comet (Szegö et al. 2000).Alfvèn (1981) showed that the interplanetary magnetic field is frozen inside the conductive plasma of the solar wind. Hence, changes in the flow by mass loading are accompanied by alterations of the magnetic field. As the modified solar wind flows around the nucleus, the magnetic field piles up ahead of the comet and gets draped by the changed flow, which forms the plasma tail (Israelevich and Ershkovich 1994, Koenders et al. 2016).

Fluorescence of incorporated ions, primarily CO^+, exited by solar radiation then causes the characteristically blue tail observable from earth. Based on a theoretical study, Wu and Davidson (1972) introduced the idea, that the interaction of the solar wind with newborn cometary ions could cause an instability in the plasma, which could in turn excite

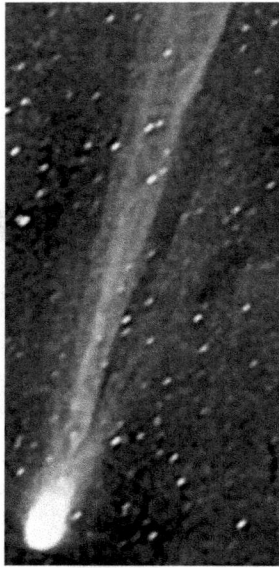

Figure 3.1: Photograph of comet 109P/Swift-Tuttle taken in 1892 by E. E. Barnard. The bright dust tail can clearly be identified above the dim plasma tail (this image is in public domain due to its age).

electromagnetic waves. The authors proposed, that these electromagnetic fluctuations are created by unstable phase space distributions in form of ion ring-beam instabilities. Later Motschmann and Glassmeier (1993) introduced non-gyrotropic phase space driven instabilities as a further mechanism to trigger electromagnetic waves.

These predictions were experimentally supported by magnetic field measurements carried out by the first spacecraft to visit a comet, the International Cometary Explorer (ICE, initially designated International Sun-Earth Explorer-3). In 1985 ICE passed through the tail of comet 21P/Giacobini-Zinner (Smith et al. 1986) at a distance of ~ 7800 km from the nucleus. The observed magnetic field showed evidence of large-amplitude plasma waves and strong turbulence significantly different from undisturbed solar wind conditions (Tsurutani and Smith 1986). Subsequent flybys at comets 1P/Halley and 26P/Grigg–Skjellerup by the Giotto and Vega spacecrafts (Glassmeier et al. 1989, Glassmeier and Neubauer 1993, Volwerk et al. 2014, Glassmeier 2017) and 19P/Borelly by Deep Space 1 (Richter et al. 2011) provided further in situ observations to support the initial idea of Wu and Davidson (1972).

The nature of the interaction between the solar wind and comets is governed by the outgassing rate of the nucleus, which strongly depends on the heliocentric distance. For highly active comets close to the Sun, enough material is sublimating to decelerate the solar wind by mass loading to the point where a cometary magnetosphere forms (Nilsson et al. 2015). One of the characteristics of such a magnetosphere is the bow shock, that forms along the stagnation streamline between the Sun and the comet as described by Glassmeier (2017). Up until the Rosetta mission, all other comets were studied close to the Sun with high production rates causing strong activity in the plasma environment. As explained by Nilsson et al. (2015), Rosettas arrival at 67P at a heliocentric distance of more than 3.6 AU made it possible to observe how the activity of the comet increased as it got closer to the Sun and perihelion, ultimately leading to the development of a cometary magnetosphere. Hence, it was possible to observe the plasma environment before typically expected features like the magnetic pileup or a bow shock (Koenders et al. 2013, 2016) formed and study the creation process of these features. As the comet got closer to the Sun, it was possible to observe the evolution of the magnetic field around 67P. From the nearly undisturbed solar wind conditions during arrival of Rosetta, to the detection of low frequency waves up to the development of diamagnetic cavities (Goetz et al. 2016, Behar et al. 2017) and steepened wave structures (Stenberg Wieser et al. 2017, Hajra et al. 2018), it was possible to combine measurements from the instruments of the RPC consortium to study the cometary interaction region.

3.2 Low Frequency Magnetic Waves

Shortly after Rosetta arrived at 67P at a heliocentric distance of 3.6 AU, low frequency quasi-coherent, large-amplitude magnetic waves below 50 mHz were detected by the RPC-MAG magnetometer. An initial analysis of these wave structures was performed by Richter et al. (2015). A typical example of the spectral characteristics of these waves is given in Fig. 3.2 using the power spectral density (PSD) for two 1024 s long intervals of typical RPC-MAG magnetic field observations. While no wave activity was detectable directly after arrival on August 2, 2014 (panel a in Fig. 3.2), one month later on September 1, 2014 two distinct spectral peaks between 20 mHz and 30 mHz emerged (panel b). During this time the cometocentric distance decreased from 456 km to 51 km, while the heliocentric distance decreased from 3.62 AU to 3.45 AU

As described in section 3.1, similar low frequency waves have been detected near comets before. Most of these waves were attributed to the pick up of the cometary ions, primarily H_2O^+ (Glassmeier et al. 1997) or proton cyclotron waves (Mazelle and Neubauer 1993). The frequencies of these waves are governed by the gyro-frequency, which is given by

$$f_g = \frac{|q|B}{2\pi m} \tag{3.1}$$

where q and m are the charge and mass of the ions and B is the strength of the background magnetic field. In the case of comet 1P/Halley waves triggered by ring-beam instabilities caused by the pick up of water group ions were detected at frequencies of 7 mHz (Glassmeier et al. 1989). Higher frequency harmonics at 21 mHz, 29 mHz and 35 mHz pumped by the water group cyclotron waves have also been observed (Goldstein et al. 1990a).

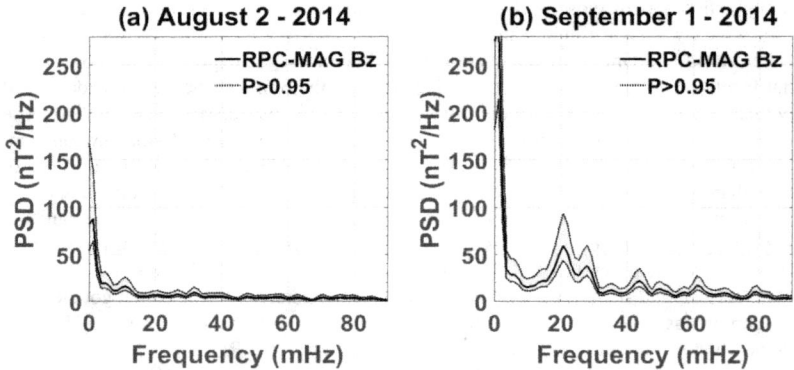

Figure 3.2: Power spectral density (PSD) estimates for two 1024 s intervals of RPC-MAG observations starting at 18:00:00 UTC on August 2, 2014 (cometocentric distance 456 km) and September 1, 2014 (cometocentric distance 51 km) for comparison. The shaded areas represent the upper and lower 95%-confidence bounds.

Hence, the detection of low frequency waves at 67P was in itself not unexpected. Equation 3.1 shows that the gyro-frequency depends linearly on the strength of the local background magnetic field. This was, however, not the case for the waves initially detected at 67P (Glassmeier 2017). These waves were initially detected in a typical range of 15 mHz up to 40 mHz (Richter et al. 2015). Even considering local variations in the magnetic field strength, this frequency range is significantly above the expected upper limit of 3.2 mHz for the water group gyro-frequency under the conditions of 67P at a heliocentric distance > 3.4 AU (Richter et al. 2015). Therefore Richter et al. (2015) suggested that a different mechanism, possibly a cross-field current instability, is responsible for the excitation of this kind of new "singing comet" waves. A cross-field current instability has been discussed in the past for example by Sauer et al. (1998) in the context of the martian magnetosphere. Based on these results, Meier et al. (2016) proposed a modified ion-Weibel instability as possible cause using a theoretical model.

3.2.1   Two-Point Low Frequency Wave Observations

Initial analysis of the singing comet waves by Richter et al. (2015) was limited by the fact, that for a given time, only observations at a single point in space were available, which made it impossible to derive for example the direction of wave propagation. The concurrent operation of RPC-MAG and ROMAP during the descent, landing and FSS of Philae were the first time the magnetic field inside the cometary plasma environment was observed by two spatially separated spacecrafts. While Richter et al. (2016) already analyzed the magnetic activity during the descent and rebound period, the following analysis focuses on the roughly 14 h of FSS observations obtained after Philae was stationary at its final landing site from November 12, 17:31:32 UTC until November 13, 14:36:39 UTC (Heinisch et al. 2017b).

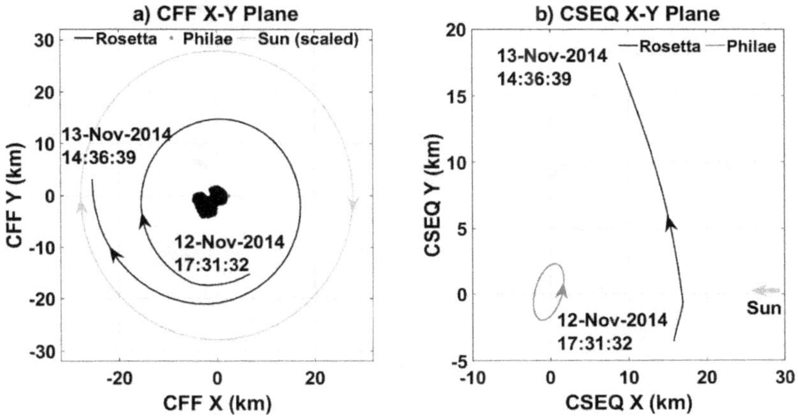

Figure 3.3: Orbit of Rosetta (black) and the Sun (scaled) relative to comet 67P from November 12, 17:31:32 UTC until November 13, 14:36:39 UTC using both the comet fixed and CSEQ (X-Y projection) frames. The arrows indicate the direction of movement (based on Heinisch et al. 2017b).

The ROMAP boom was deployed shortly after separation, therefore influence from lander systems on the observations was minimal. ROMAP was operated at a constant 1 Hz sampling rate. Because of operational constraints the instrument had to be switched off on November 13, from 04:05:34 UTC until 06:31:07 UTC, therefore no measurements are available during this period. Until 18:01 UTC RPC-MAG was operated in its 20 Hz burst mode and was switched into its 1 Hz normal mode for the remainder of the FSS. Hence, the first part of the orbiter observational data was downsampled to 1 Hz. For the following analyses only measurements from the RPC-MAG outboard magnetometer were taken into account, while the observations from the inboard magnetometer were only used for cross checking and calibration purposes. To be of use for any further analysis, all ROMAP magnetometer measurements had to be rotated into a known global (cometary) reference frame. In this case a Comet Centered Solar Equatorial (CSEQ) coordinate system was used. The center of this reference frame is the center of mass of the comet, the positive x-axis points towards the Sun, the positive z-axis is the component of the Sun's north pole of date orthogonal to the positive x-axis and the y-axis completes the right handed system. This reference frame is used for all analyses in this chapter. For all reference system conversions the NASA NAIF SPICE (Acton 1996) system and the lander attitude provided by Heinisch et al. (2016) were utilized.

The orbit of Rosetta relative to Philae and comet 67P between November 12, 17:31:32 UTC and November 13, 14:36:39 UTC is illustrated in Fig. 3.3 using both the comet fixed frame (CFF) and the CSEQ frame. The center of the CFF is the center of mass of the comet, the z-axis is the positive pole of the rotation axis, the x-axis is defined by the

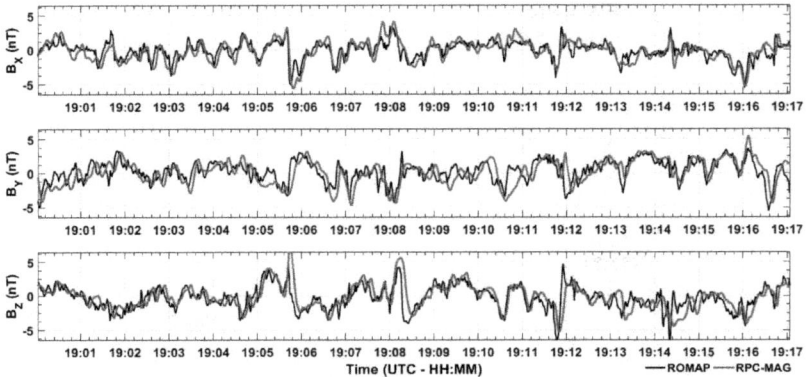

Figure 3.4: Calibrated ROMAP (black) and RPC-MAG (red) FSS dataset (CSEQ coordinate system) from November 12, 19:00:00 UTC until November 12, 19:17:04 UTC (1024 s duration) at a distance between Rosetta and Philae of 18.8 km (based on Heinisch et al. 2017b).

intersection point between equator and prime meridian and the y-axis completes the right handed system. Rosetta remained in the same terminator orbit and the distance between Philae and the orbiter stayed roughly constant at around 18.5 km. This distance was accurately constrained by repeated CONSERT (Kofman et al. 2007) ranging measurements during FSS intended to help localize Philae.

A 1024 s long interval of both ROMAP and RPC-MAG observations is displayed as a typical example in Fig. 3.4. The magnetic field variations measured on the surface by ROMAP are nearly identical to the RPC-MAG observations in orbit (at 18.8 km distance). This is true for the entire FSS interval, as was already discussed by Heinisch et al. (2016). This implies that no large-scale boundary layer exists above the cometary surface that is influencing the magnetic field variations. The observations are dominated by low frequency waves, with typical amplitudes in the range of ± 5 nT (Fig. 3.4). Richter et al. (2015), Richter et al. (2016) and Glassmeier (2017) already described these waves, which were first detected after the comet rendezvous around August 6, 2014. The mean Pearson correlation coefficient (described i.e. by Benesty et al. 2009) ρ for all three components for the entire FSS interval is 0.86 after accounting for a wave propagation delay of 5 s. Therefore, the nucleus itself has no impact on the magnetic field measurements and no large-scale surface boundary layer exists, which agrees with previous conclusions of Richter et al. (2016) and Heinisch et al. (2016). The simultaneous operation of two spatially separated magnetometers in the same plasma environment opens up the possibility to study plasma waves in much higher detail. Assuming plane waves, one can determine the wave velocities and the orientation of the wave-vector $\vec{k}$. This was accomplished by combining minimum variance analysis (MVA) as explained by Sonnerup and Cahill (1967) and Sonnerup and Scheible (1998) with a cross-spectral analysis. The MVA can be used to determine the direction of the wave k-vector (Smith and Tsurutani 1976, Verkhoglyadova et al. 2013) but is ambiguous as to the sign. To constrain the orienta-

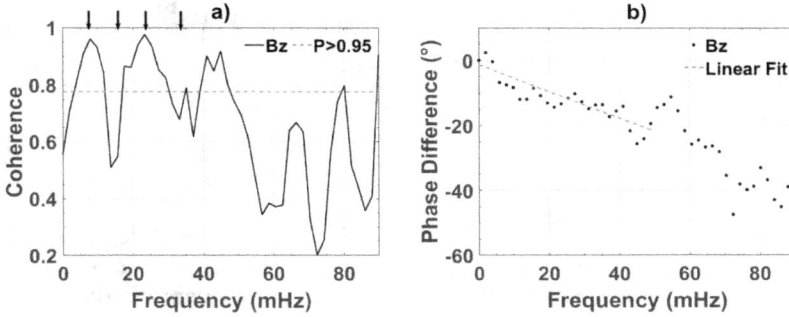

Figure 3.5: Typical squared coherence (a) and phase shift (b) between the z-components of the ROMAP and RPC-MAG signals shown in Fig. 3.4 from 19:00:00 UTC until 19:15:00 UTC on November 12 at a distance of 18.8 km. Most of the coherent waves are below 50 mHz. The dashed line at 0.77 is the statistical significance threshold and the arrows indicate peaks in the coherence correlating to peaks in spectral density (Fig. 3.6). The negative phase relation in this frequency range indicates, that the wave signatures are first observed by ROMAP (based on Heinisch et al. 2017b).

tion of the projected k-vector, the phase shift gained from cross-spectral analysis can be used, assuming plane waves propagating along the projected k-vector direction with a finite speed. Even though the MVA has limitations when used to classify discontinuities, especially with noisy data as investigated by Hausman et al. (2004), it is perfectly suited to this kind of analysis, due to the periodic nature of the waves and the signal to noise ratio.

3.2.2 Spectral Analysis

The squared coherence and corresponding phase relation was calculated between the individual components of the ROMAP and RPC-MAG observations based on the method proposed by Welch (1967). The results for the Bz-components are plotted in Fig. 3.5, using the same interval as in Fig. 3.4, as a typical example. Most of the coherent waves above the significance threshold of 0.77 (calculated based on Shumway and Stoffer (2011) for the used six degrees of freedom) are below 50 mHz, which is consistent with the previous observations during descent and landing (Richter et al. 2016) and typical for the singing comet (Richter et al. 2015). It also matches the expected frequency range for a modified ion-Weibel instability as discussed by Meier et al. (2016) and Glassmeier (2017). For this frequency range, the phase between RPC-MAG and ROMAP decreases approximately linear with increasing frequency. This behaviour can be observed in all three components independent of the selected time interval, which indicates that the waves in the magnetic field are always first observed by the ROMAP sensor and then propagate outwards reaching the RPC-MAG magnetometer onboard the orbiter. Assuming a plane wave, the slope of this curve is proportional to $2\pi s/v_p$, where s is the distance the wave travelled and v_p is the phase velocity. The nearly linear behaviour suggests, that the velocity v_p (in this

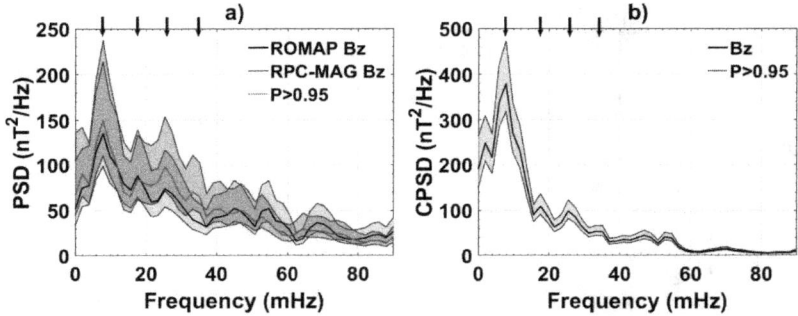

Figure 3.6: Power spectral density (PSD) estimates of ROMAP and RPC-MAG (a) and estimated cross power spectral density (CPSD) (b) for the interval displayed in Fig. 3.4 from 19:00:00 UTC until 19:15:00 UTC on November 12. The shaded areas represent the upper and lower 95%-confidence bounds. In this example most of the power in both signals is at ≈ 10 mHz. The arrows indicate spectral peaks correlating to peaks in coherence (Fig. 3.5) (based on Heinisch et al. 2017b).

example $v_p \approx 5.3$ km/s) is roughly constant for the frequency range considered. Based on this information it is not only possible to determine v_p but a corresponding constraint can also be used with the MVA to determine the orientation of the k-vector. While the coherence (Fig. 3.5) shows five significant peaks at ~8 mHz, ~16 mHz, ~24 mHz, ~32 mHz and ~40 mHz, the power spectral densities (PSD) and cross power spectral density (CPSD) in Fig. 3.6 only exhibit peaks at ~8 mHz, ~16 mHz, ~24 mHz, ~32 mHz (marked by arrows in the individual figures). These peaks hint at a harmonic excitation of waves similar to what was previously observed at comet 1P/Halley (Glassmeier et al. 1989, Goldstein et al. 1990b). These patterns are present at different times during the entire observation, but with shifted frequency ranges, which is to be expected as the main singing comet frequency fluctuates over time. Different types of window-functions have been used to calculate the PSD, to compare the results and exclude systematic effects.

The spectral analyses described above where applied to the entire dataset, by dividing it into 1910 smaller intervals of 1024 s length overlapping each other by 480 s. The spectra were estimated based on the method proposed by Welch (1967). Fig. 3.7 displays the dynamic coherence spectrum (top) and dynamic cross power spectral density (bottom) for the Bz-component of the entire observation interval from November 12, 17:31:32 UTC until November 13, 14:36:39 UTC as an example. During the entire time coherent waves corresponding to significant power can not only be observed at a single frequency, but are spread out over the entire range between 5 mHz and 50 mHz with a relatively sharp cutoff. The fluctuations in the cross power density are not linked to diurnal (12 h or 6 h period) variations and must therefore be caused by changes in the plasma environment.

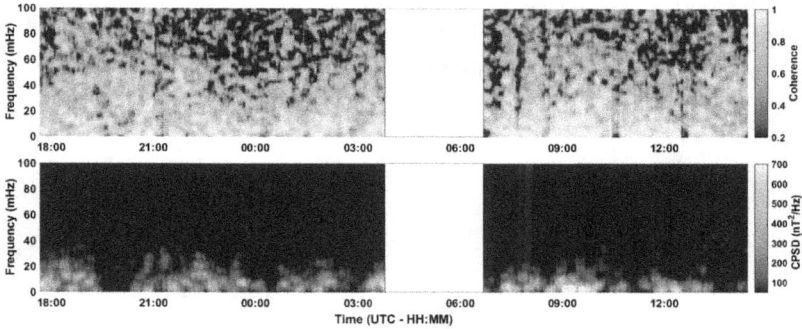

Figure 3.7: Dynamic coherence spectrum (top) and cross power spectral density (bottom) for the Bz-component of the entire interval from November 12, 17:31 UTC until November 13, 14:37 UTC. The significance threshold for the coherence (calculated based on Shumway and Stoffer (2011) for the used 6 degrees of freedom) is at 0.77 (green) (based on Heinisch et al. 2017b).

3.2.2.1 Wave Evolution

The frequency of the waves observed during the FSS spans from 5 mHz to 50 mHz with an average frequency in the range of 8 mHz. While this range is in general consistent with the observations by Richter et al. (2015, 2016), the most significant frequency of approximately 8 mHz observed during the FSS is below the typical singing comet frequency range identified initially by Richter et al. (2015). It is much closer to the 3.2 mHz expected for the water group gyro-frequency (Richter et al. 2015). A decline in frequency over time was already observed by Richter et al. (2016) based on the magnetic two-point observations during the 7 h descent of Philae. This frequency change continued during the FSS. Because of the simultaneous change in the distance between Philae and Rosetta and the comet, Richter et al. (2016) were not able to attribute this frequency shift to a single cause. Interference from the Philae flywheel (described in section 4.1) in the frequency range of the singing comet waves might also have influenced initial cross spectral analysis performed by Richter et al. (2016).

As the PSDs for the observations by RPC-MAG and ROMAP are almost identical during the SDL and FSS (see e.g. Fig. 3.6), the change in mean wave frequency is not caused by observational bias due to the spatial separation between the orbiter and lander. Hence, at this outgassing level, close to the nucleus between 15 km and 30 km, the cometocentric distance has very limited influence on the spectral characteristics of the observed waves.

During the progression of the Rosetta mission the spectral characteristic of the magnetic field changed continuously. While the field was initially dominated by singing comets waves (see Fig. 3.2), these distinct low frequency variations vanished in the increasingly turbulent field observed closer to perihelion. As an example the PSD for an 1024 s interval of magnetic field observations from February 22, 2015 at a heliocentric distance

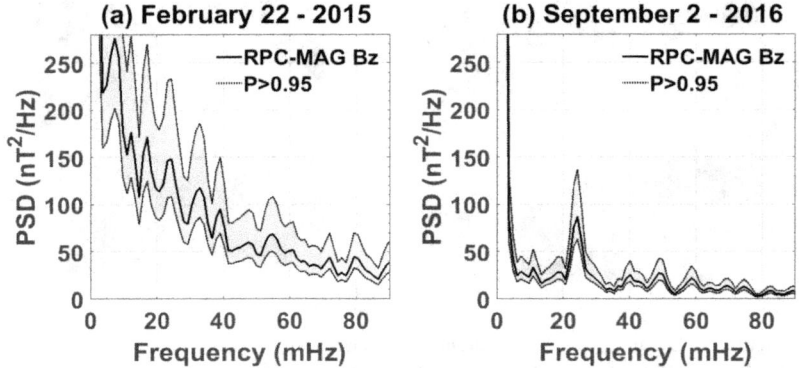

Figure 3.8: Power spectral density estimates for two 1024 s intervals of RPC-MAG observations starting at 18:00:00 UTC on August 2 (cometocentric distance 55 km) and September 1 (cometocentric distance 24 km) for comparison. The shaded areas represent the upper and lower 95%-confidence bounds.

of 2.25 AU and a cometocentric distance of 55 km is depicted in panel a) of Fig. 3.8. Shortly after, around March 2015, the magnetic environment started to change due to the increasing activity of the comet. Steepened magnetic waves (Stenberg Wieser et al. 2017, Hajra et al. 2018), similar to what was already observed at 21P/Giacobini-Zinner (Tsurutani et al. 1987) and diamagnetic cavity structures (Goetz et al. 2016) started to emerge and become increasingly prevalent, while the comet got more active. After perihelion on the outbound trajectory away from the Sun, these phenomena subsided and at a heliocentric distance of approximately 3.30 AU around July 2016 singing comet waves again became a typical feature of the observed magnetic field. As an example the PSD is shown for an interval from July 5, 2016 at a heliocentric distance of 3.30 AU and a cometocentric distance of 24 km in panel b) of Fig. 3.8. At this time the low frequency waves exhibit the same spectral properties as the waves observed in late 2014, and the magnetic activity has returned to almost the same state. The PSDs in Fig. 3.2 and Fig. 3.8 illustrate, that the spectral characteristics of the observed magnetic field are not just governed directly by the heliocentric distance of the comet. Observations during the descent of Philae already showed, that small scale changes in the cometocentric distance have little influence on the observed structure of the magnetic field, as the typical plasma scales are much larger. Variations in the magnetic activity can rather be linked to changes in the outgassing rate of the comet (Hansen et al. 2016), which depends on the heliocentric distance, but is not completely symmetric around perihelion (Snodgrass et al. 2013). As the changes in the magnetic field are linked to variations in the plasma environment, the local cometary neutral and plasma densities can help understand the spectral characteristics of the magnetic field. An example is given in Fig. 3.9, depicting the frequency of highest power in the singing comet range, the corresponding power density and the neutral pressure observed by the ROSINA-COPS neutral gauge (Balsiger et al. 2007). The neutral pressure is directly proportional to the neutral density, but the calibration parameters can

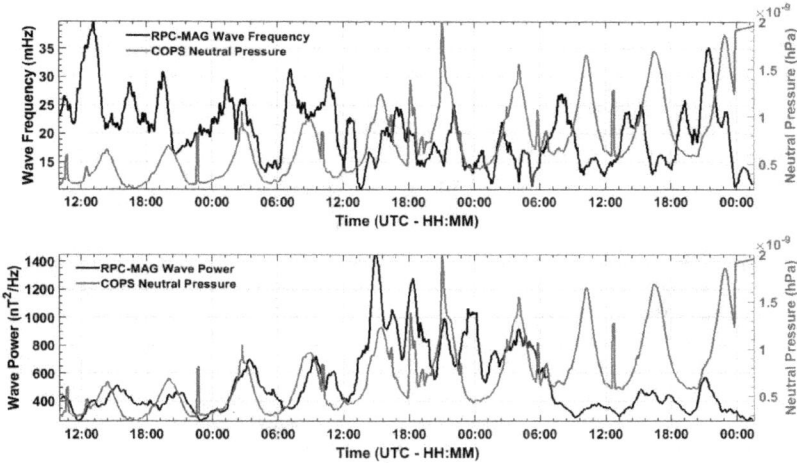

Figure 3.9: Most significant magnetic wave frequency and corresponding power based on RPC-MAG observations in comparison to the ROSINA-COPS neutral pressure observed between November 6 and 9, 2014 at a cometocentric distance of 30 km.

slowly change over time and depend on the instrument mode (Balsiger et al. 2007). As no officially calibrated density information is available, the directly observed pressure was used for comparison. An interval between November 6 and 9, 2014 with the orbiter at a cometocentric distance of ~ 30 km was chosen as example. This interval was before lander pre-separation operations and orbital corrections caused possible interference, making later measurements at similar distances unsuitable for analysis. These observations show, that the variations in wave power directly correlate ($\rho = 0.73$) with changes in the neutral pressure. Variations are predominantly in phase, but with temporary deviations up to 30°. The most significant frequency, which was determined based on the highest power in the singing comet frequency range below 50 mHz, is not directly linked to the neutral pressure. A cross correlation analysis shows a varying phase shift in the range of 90° to 180° between frequency and pressure. Even considering these phase shifts, the correlation coefficient never exceeds 0.61. Although the general behaviour qualitatively matches the predictions of Meier et al. (2016) for a modified ion-Weibel instability, it would require a very narrow range for the water ion density to achieve the observed phase shift between frequency and power. Because of the complex nature of the plasma environment close to the nucleus, numerical simulations would be required, to use the observational results to validate possible mechanisms behind the wave activity and further analyze the underlying process.

A spectral analysis revealed, that both the neutral pressure and the wave power exhibit a 3.1 h, 6.7 h and 13.1 h periodicity. As these frequencies are directly linked to the comet rotation rate, the variations are most likely linked to the diurnal changes in the activity of

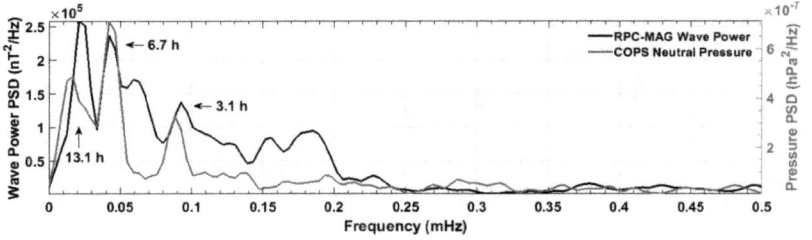

Figure 3.10: Power spectral density of the ROSINA-COPS neutral pressure observed between November 6 and 9, 2014 in comparison to the Power spectral density of the magnetic wave power. Multiples of the comet rotation period at 3.1 h, 6.7 h and 13.1 h are marked.

67P (La Forgia et al. 2017). With increasing cometary activity and larger cometocentric distances of Rosetta, the observed correlation diminishes.

3.2.3   Wave Orientation and Phase Velocity

3.2.3.1   Wave Orientation

To determine the orientation of the k-vector smaller intervals of only 256 s overlapping by 224 s were used as input for the MVA to select individual wave packets and reduce the influence of averaging. In a first step, an 4th order Butterworth lowpass filter (Butterworth 1930) with a cutoff at 50 mHz was applied, to exclude higher frequencies without significant power. In a second step, an 4th order Butterworth bandpass filter with an bandwidth of 10 mHz centered around the most significant frequencies derived in section 3.2.2 (8 mHz, 16 mHz, 24 mHz, 32 mHz) was used, to study a potential frequency dependence of the k-vector.

The above mentioned CSEQ frame was used for both input datasets to be able to directly relate the results to the nucleus as a common point of reference. In accordance with Richter et al. (2016) only MVA results with a corresponding eigenvalue ratio of $\lambda_{med}/\lambda_{min} \geq 3$ were considered for further analysis. This way intervals without a distinct wave pattern were excluded, while still keeping enough events to be of statistic significance. Additionally the mean maximum cross correlation was calculated between all three corresponding components. Only intervals with a correlation $\rho > 0.7$ were used, to ensure that strong external disturbances (i.e. instrument heaters) were excluded, while still accounting for differences in the individual waves. The histogram in Fig. 3.11 shows the overall abundance of the resulting angles between the wave-vector direction for the individual wave packets and the Sun for both ROMAP and RPC-MAG projected onto the X-Y plane and X-Z plane respectively. The histogram shows, that the angle distribution is very similar for ROMAP and RPC-MAG as already observed during descent (Richter et al. 2016).

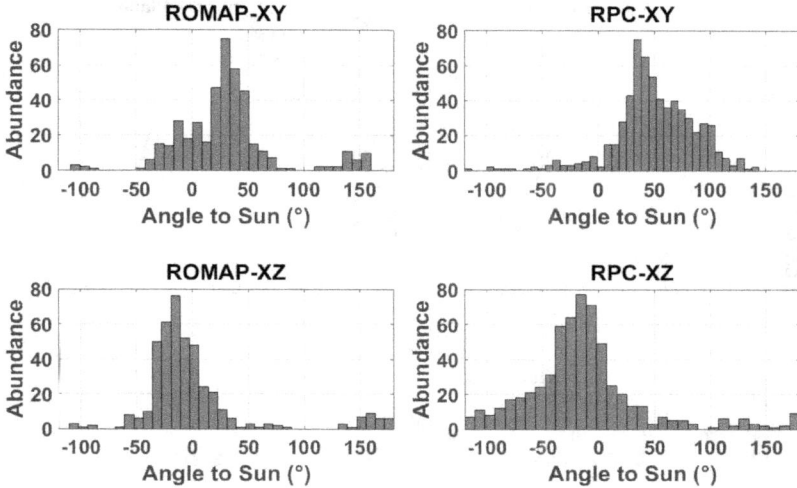

Figure 3.11: Abundance of the resulting angles between the wave-vector direction for the individual wave packets as determined by the MVA and the Sun for both ROMAP and RPC-MAG projected onto the X-Y plane and X-Z plane respectively (adopted from Heinisch et al. 2017b).

Most of the waves propagate roughly against the solar wind direction, as was already theoretically predicted by Meier et al. (2016). While the waves in the x-z plane are centered around -15° relative to the Sun, the results for the x-y plane are more complex. Both instruments show that most of the waves are centered around 50° relative to the Sun, but while the ROMAP results show a considerable amount of waves centered around 0°, the RPC-MAG results show a second population of waves centered around 100°.

Applying a bandpass filter to select only the most significant frequencies, as determined in section 3.2.2, only causes minor changes in the derived wave-vector direction. Compared to the overall analysis using only a lowpass filter with an upper cutoff of 50 mHz, the angular deviation is at maximum 10°. To visualize the overall geometry of the observations, the orientations of the individual magnetic wave packets are shown in Fig. 3.12 as unit vectors projected onto the X-Y and X-Z planes originating from the orbiter trajectory or comet center respectively. To preserve visibility each vector in Fig. 3.12 represents the average of ten consecutive non overlapping results. The temporal evolution is illustrated by color coding both sets of vectors, starting with blue indicating November 12. 17:34:44 UTC and transitioning to red indicating November 13. 14:15:10 UTC. No diurnal variation or dependence on the spacecraft trajectory or comet rotation (comet rotation period is 12.5 h) can be observed. The variation in the wave orientation increases significantly between approximately 7:00 UTC and 11:00 UTC on November 13. The orientation of the waves during this interval matches with the deviations in Fig. 3.11. As there are no sudden changes in spacecraft operations during this interval, this change can not be at-

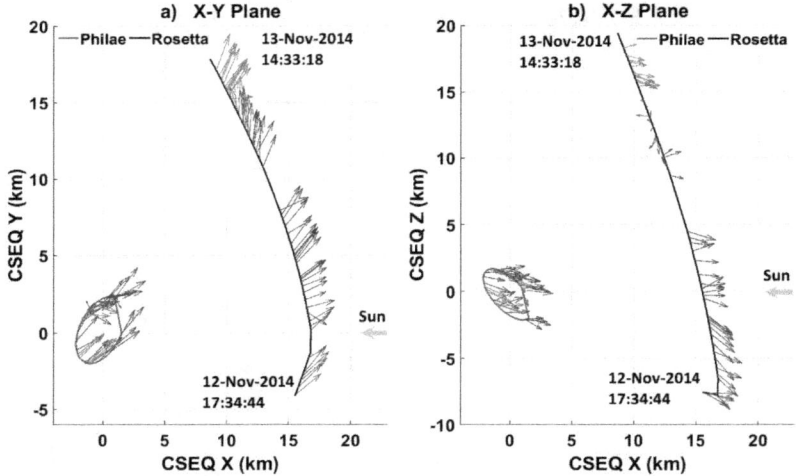

Figure 3.12: Orientation of the wave-vectors based on MVA for ROMAP (following Philae trajectory) and RPC-MAG (following the orbiter trajectory) projected onto the X-Y plane (panel a) and X-Z plane (panel b). To preserve visibility each vector represents the average of ten consecutive non overlapping result. The color coding illustrates the temporal evolution, starting with blue for November 12. 17:34:44 UTC and transitioning to red indicating the end of the FSS on November 13. 14:15:10 UTC. The Sun is always at the right hand side of the panels as indicated by the yellow arrow (based on Heinisch et al. 2017b).

tributed to artificial interference or attitude changes. This phenomenon can also not be caused by periodic changes in comet activity, because if this was the case, it should have been observed earlier or even during the descent (Richter et al. 2016).

3.2.3.2   Phase Velocity

Based on the nearly identical wave propagation direction for RPC-MAG and ROMAP, the projected phase velocity of the individual waves was calculated using the phase difference. The slope $\Delta\varphi/\Delta f$ of the phase relation is given by $2\pi s/v_p$ where s is the distance the wave traveled. This distance s can be calculated using the normalized wave propagation direction $\vec{k}$ obtained with the MVA and the normalized Rosetta position vector $\vec{p_{ROS}}$ and the orbiter lander distance d: $s = (\vec{k} \cdot \vec{p}_{ROS})^{-1} d$. The relative temporal uncertainty between lander and orbiter observations is below 1 s. The error in the relative distance is expected to be below 50 m (constrained by several CONSERT ranging measurements) and hence, has a negligible effect on the results. To reduce possible errors, only waves with a difference of less that $45°$ in the propagation angle relative to the Sun have been included. A histogram of the resulting velocities is displayed in Fig. 3.13. The mean phase velocity based on this analysis is 5.3 km/s with a standard deviation of 1.4 km/s. These results are consistent with the ones described by Richter et al. (2016) for the earlier SDL

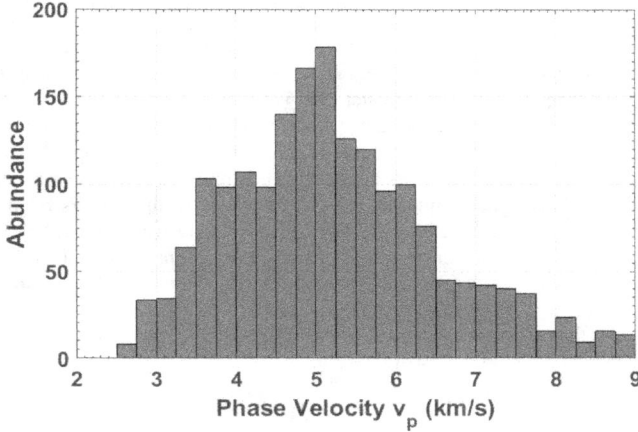

Figure 3.13: Abundance of the phase velocities estimated for individual intervals based on the phase difference and propagation direction. The mean velocity is 5.3 km/s with a standard deviation of 1.4 km/s (adopted from Heinisch et al. 2017b).

observations. With the known phase velocities the projected wavenumbers $k_p = 2\pi f / v_p$ and projected wavelengths $\lambda_p = 2\pi / k_p$ can be calculated. At an average wave frequency of about $f = 8$ mHz and assuming a mean velocity of $\overline{v_p} = 5.3$ km/s one gets a wavenumber of $k_p \approx 9.5 \cdot 10^{-3}$ km^{-1} and a wavelength of $\lambda_p \approx 662$ km, which agrees very well with theoretical predictions discussed by Meier et al. (2016) and Glassmeier (2017).

3.3   Magnetic Field Attitude Determination

Singing comet waves are not only of scientific interest, but also opened up the possibility to determine the attitude of the Philae lander in flight and on the surface of CG. Compared to a compass on earth, which uses the strong and quasi constant planetary dipole field to orient the needle, the magnetic field in space is much more intricate. Especially around comets with the solar wind interaction, the magnetic field close to the nucleus becomes dominated by waves and turbulence, as described in the previous chapter. In the absence of strong large scale planetary fields, known coherent magnetic structures need to be present to allow magnetic orientation. This becomes especially true, if measurements are also influenced by temperature dependent sensor offsets or bias fields caused for example by spacecraft components. In this case only the time varying components of the magnetic field can be used. This can even be the case if high accuracy attitude determination is necessary on earth and small scale changes in the constant field become important (as described by Heinisch and Auster 2015).

In these cases a reference sensor with known orientation becomes necessary and the magnetic field fluctuations need to be coherent on spatial scales larger than the distance between the reference magnetometer and the target sensor with unknown orientation. While these prerequisites are not necessarily met within the undisturbed solar wind at heliocentric distances around 3 AU, the singing comet waves completely fulfilled those requirements. Hence, Heinisch and Auster (2015) developed a high accuracy method to determine the attitude of a magnetometer relative to a reference sensor using magnetic field fluctuations. The idea behind this approach is to compare the orientation of time dependent 3D magnetic field vectors observed concurrently by multiple magnetometers at different locations to determine the orientation of one of the sensors relative to the other. The basic algorithm was developed by Heinisch and Auster (2015) and initially tested with ground based magnetic observatory measurements. Further improvements were than made by Heinisch et al. (2016) to adapt this method to the Philae landing.

To reconstruct the orientation of one of the sensors, the observed magnetic field vectors are rotated using a rotation matrix of the form:

$$\underline{\underline{M}} = \begin{pmatrix} 1 & 0 & 0 \\ 0 & \cos(\alpha) & -\sin(\alpha) \\ 0 & \sin(\alpha) & \cos(\alpha) \end{pmatrix} \begin{pmatrix} \cos(\beta) & 0 & \sin(\beta) \\ 0 & 1 & 0 \\ -\sin(\beta) & 0 & \cos(\beta) \end{pmatrix}$$
$$\begin{pmatrix} \cos(\gamma) & -\sin(\gamma) & 0 \\ \sin(\gamma) & \cos(\gamma) & 0 \\ 0 & 0 & 1 \end{pmatrix} \tag{3.2}$$

with the Euler angles α, β and γ. For each set of euler angles, the mean Pearson correlation coefficient between the rotated three component signal and the measurements from one or multiple reference magnetometers is calculated. The rotation angles corresponding to the highest correlation then describe the attitude of the target magnetometer relative to the reference sensor. It would be possible to solve this problem with an optimization method instead of exhaustively trying every possible combination of Euler angles for a given angular resolution. But as the chosen approach eliminates the possibility of incorrectly resulting in a local instead of a global correlation maximum, it offers a significant advantage. Having information for all combination of angles available, also makes it possible to gauge the sensitivity of the result to small angular changes, which gives an estimate for the lowest achievable angular resolution of this method. As each of the necessary calculations is computationally completely independent, it can be implemented very effectively on modern parallel processors. The different steps of the method as explained in detail by Heinisch et al. (2016) are illustrated in the flowchart in Fig. 3.14.

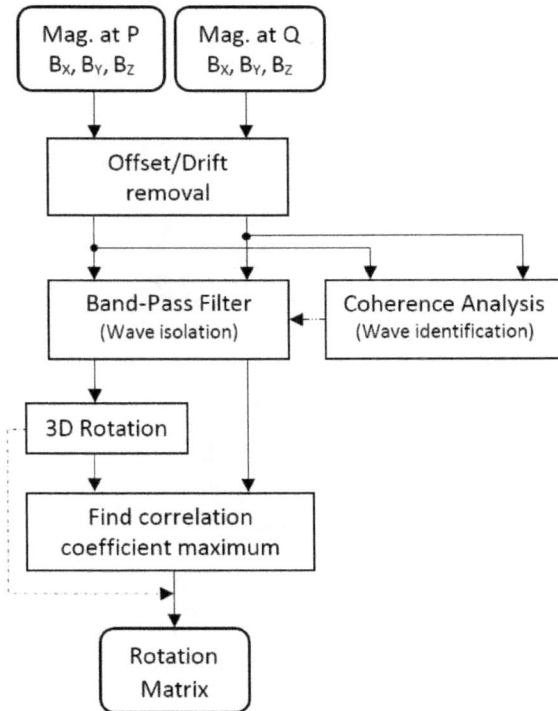

Figure 3.14: Flowchart illustrating the different steps used for magnetic field attitude determination based on Heinisch et al. (2016).

4 Philae Flight Reconstruction

Because of technical constraints, Philae was not equipped with dedicated navigation instruments, as described in section 2.5.1. Hence, observations from scientific instruments had to be combined with technical housekeeping measurements to reconstruct the attitude of the lander. Even before the arrival of Rosetta at the comet, different methods were investigated for this purpose. Initially the power output of the different solar arrays of Philae, (see Fig. 2.8), combined with images taken from orbit and by the lander were intended as primary way to establish the attitude of Philae (Remetean et al. 2016). The total output current of each panel depending on insolation is given by

$$I_{tot} \approx k \cos(\alpha) \tag{4.1}$$

where α is the solar incidence angle for each of the modules and the factor k describes the properties of the module. Ideally this factor should be constant for a given solar panel. Because of variations in the electrical components and technical limitations of the power converters and maximum power point tracking circuits, this value was in reality a function of the raw solar array output voltage. As the solar array housekeeping measurements were initially not intended for this kind of analysis, no detailed ground characterization was performed. This limited the accuracy of this potential method significantly. The maximum sampling rate of 4 mHz additionally hampered a dynamic analysis of the array power. Especially with the risk of a not nominal landing, due to possible problems with the harpoons and uncertainties with the MSS, it became clear that a more precise attitude determination method was needed. This situation was compounded by the unexpected shape of the comet. The irregular nature made it very difficult to accurately predict illumination conditions and possible safe landing sites, which was not expected based on the shape initially proposed by Lamy et al. (2006) based on a lightcurve analysis performed with Hubble images. Additional problems were encountered while generating coordinate systems describing 67P, especially defining longitude and latitude, as the concavity lead to coordinate ambiguities (Preusker et al. 2015).

Hence, it was decided to use two-point magnetic field measurements as primary source for attitude information. This approach as described in section 3.3 uses 3D magnetic field observations to reconstruct the orientation of a target magnetometer relative to a reference magnetometer with known orientation. To achieve this, the magnetic field vectors observed by the target magnetometer are correlated with the vectors observed by the reference magnetometer to determine the rotation matrix (i. e. eq. 4.2) between the coordinate systems of the two sensors. In this case RPC-MAG on the orbiter was used as reference with known orientation, while ROMAP was the target magnetometer. This method was successfully used to determine the final orientation of Philae as described by Heinisch et al. (2016) and Jurado et al. (2016) and the attitude during descent and rebound as explained below.

4.1 Descent Overview

On November 12, 2014 at 08:35 UTC Philae was separated from the Rosetta spacecraft, descending on a free ballistic trajectory towards the surface of comet 67P. The actual separation was nominal and has already been described in detail (Auster et al. 2015, Biele et al. 2015, Garmier et al. 2015, Jurado et al. 2016, Baranyai et al. 2016).

Philae was actively stabilized by an internal flywheel during descent. The rotation axis of this flywheel was approximately parallel to the lander z-axis (see Fig. 2.7). This insured accurate antenna pointing and kept the landing gear aligned for touchdown, but allowed Philae to rotate roughly around its z-axis. The rotation rate changed over time due to small changes in the flywheel rotation rate. To accurately reconstruct the descent attitude, these changes in rotation have to be determined.

After separation the CONSERT radio experiment (Rogez et al. 2016) as well as the ROMAP and RPC-MAG sensors were all operating simultaneously. Though the main purpose of the CONSERT instrument operation was to track the distance between Rosetta and Philae during descent, it was also possible to reconstruct the rotation from periodic changes in the CONSERT observations. The following analysis combines the two-point magnetic analysis with housekeeping information and CONSERT radio measurements to reconstruct the descent of Philae.

CONSERT operated from separation up until 40 min before the nominal time of touchdown (Kofman et al. 2015). These measurements were performed every 2.5 s. Based on the bi-static CONSERT sounding principle (Kofman et al. 2015) it was possible to track the distance between orbiter and lander through the delay in the direct signal path of the CONSERT received signal. In this data, a modulation of the received signal power in the path between orbiter and lander could be observed (Hahnel et al. 2015). This modulation is a result of the antenna properties (frequency dependent pattern and polarization) of CONSERT lander and orbiter antennas and the movement and changes in attitude of lander and orbiter. The CONSERT results for the rotation rate are shown in Fig. 4.1. Fluctuations, especially the increase around 12:30 UTC, are likely due to the CONSERT antenna properties and caused by attitude corrections of the orbiter during the descent (Heinisch et al. 2017a).

The dynamic power spectrum of the magnetic field observed by ROMAP during descent (see Fig. 4.2) shows two distinct signatures at ~14 mHz and ~2 mHz. While the higher frequency signature was caused by the flywheel electronics, the lower one was directly caused by the lander rotation. The magnetic field created by the high frequency drive circuit of the flywheel motor can be observed in the ROMAP spectrum due to alias-

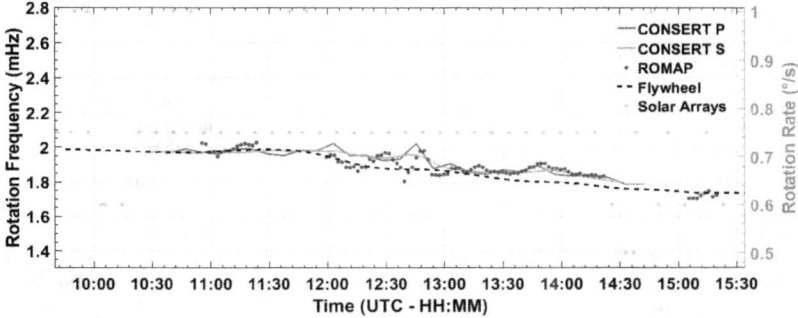

Figure 4.1: CONSERT S: Rotation rate from spectrogram of the variations of the signal power during the descent. CONSERT P: Rotation rate from peak detection. Flywheel: Scaled Flywheel rotation rate reconstructed based on ROMAP observations. Solar Arrays: Rotation rate based on solar panel (panel 2 and 4) current fluctuations. ROMAP: Rotation rate reconstructed from magnetic field measurements (based on Heinisch et al. 2017a).

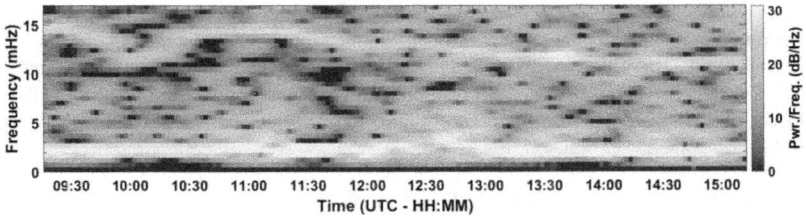

Figure 4.2: Dynamic power density spectrum of the Bx-component of the ROMAP magnetic field observations after separation from the orbiter. Two prominent signatures are visible at ~ 14 mHz, caused by the flywheel electronics and at ~ 2 mHz caused by the lander rotation (based on Heinisch et al. 2017a).

ing, even though the ROMAP sampling rate was significantly lower than the flywheel rotation rate. Ground based tests confirmed, that this aliased frequency is proportional to the actual flywheel rotation rate. Therefore, the scaled 14 mHz signature can be used to reconstruct the flywheel rotation. As this flywheel signature is in the range of the magnetic background waves, as discussed in section 3.2.2, the ROMAP observations during this part of the rebound are of limited use for scientific analysis even with perfect knowledge of the rotation and attitude.

Because the ROMAP sensor reference frame is aligned with the lander frame (the lander reference frame is shown in Fig. 2.7), a rotation around Philae's z-axis causes the observed x- and y- components of the magnetic field to change periodically. This creates a nearly sinusoidal signature in the observed magnetic field with the same frequency the lander is rotating at. As ROMAP was operating with a sampling rate of 1 Hz, the much slower rotation of Philae was accurately resolved. Fig. 4.1 shows the reconstructed rotation frequency based on the period of the sinusoidal signatures in the x- and y-components of the ROMAP observations. Additionally records from two of the solar array currents (solar arrays 2 and 4 - see Fig. 2.8) were also used to reconstruct the rotation frequency (shown in red) by analyzing the changes in the solar array output current due to the changing insolation patterns. The solar array housekeeping data was sampled every 12 s, therefore the temporal resolution is much worse in comparison to the ROMAP and CONSERT results. The interpretation of the solar array currents is furthermore complicated by geometric influences, for example due to self shadowing. Together with the setup of the lander solar arrays, this makes an unambiguous attitude reconstruction based solely on solar array currents impossible (Jurado et al. 2016). Nevertheless this comparison made it possible to analyze the accuracy and limitations of this approach, which was useful especially for the attitude reconstruction between the second and third touchdown, where no CONSERT measurements were available.

The solar array current, CONSERT and ROMAP reconstructions all exhibit a decline in lander rotation rate over time. Based on the ROMAP observations this decrease in the lander rotation frequency seems to be caused directly by a decline in the flywheel rotation frequency. This conclusion can be drawn from the fact that the relative slope of the flywheel signature (see Fig. 4.2 at approx. 14 mHz) is the same as the one observed in the rotation frequency. Ground based tests using the ground reference models confirmed, that the frequency of the magnetic signature caused by the flywheel electronics is directly linked to the rotation speed of the flywheel itself but subject to aliasing. Unfortunately as no high resolution flywheel housekeeping data is available for the descent period, no direct comparison is possible.

As the rotation frequency is already known, only the phase of this rotation, the relative alignment of the rotation-axis and the global alignment of the lander z-axis had to be determined to completely reconstruct the attitude. First the orientation of the rotation axis relative to the lander z-axis was determined, than in a second step, the phase of the rotation was reconstructed. Finally the actual orientation of Philae relative to Rosetta was determined.

To ascertain the alignment of Philae's rotation axis, the RPC-MAG timeseries was first transformed into the comet fixed CHEOPS coordinate-system (Preusker et al. 2015) and than numerically rotated to find the best match between the ROMAP and RPC-MAG z-components similar to what was described by Heinisch et al. (2016) and Heinisch and Auster (2015). The NASA NAIF SPICE system (Acton 1996) was used to facilitate the coordinate calculations. In a second step the x- and y-component of the ROMAP observations were numerically rotated (using the Euler angles α and β) against the ROMAP z-axis (which is parallel to the lander z-axis) to minimize the power $P_z(2 \text{ mHz})$ of the rotational signature in the dynamic power spectrum, while simultaneously increasing the correlation with the RPC-MAG z-observations. The rotation matrix applied to the RPC-MAG and ROMAP data was of the type:

$$\underline{\underline{M}} = \begin{pmatrix} 1 & 0 & 0 \\ 0 & \cos\alpha & -\sin\alpha \\ 0 & \sin\alpha & \cos\alpha \end{pmatrix} \begin{pmatrix} \cos\beta & 0 & \sin\beta \\ 0 & 1 & 0 \\ -\sin\beta & 0 & \cos\beta \end{pmatrix} \tag{4.2}$$

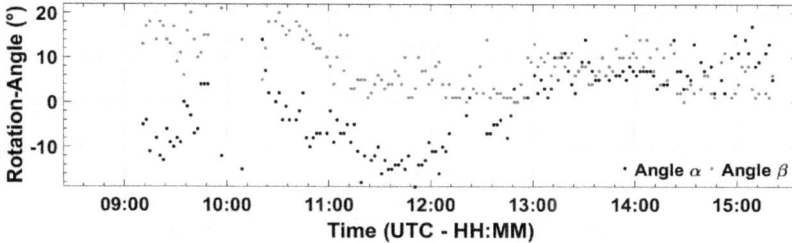

Figure 4.3: Rotation angles for the relative alignment of the lander rotation axis for individual 20 min intervals overlapping by 90% (based on Heinisch et al. 2017a).

Table 4.1: List of reconstructed descent attitude segments achieving errors better than ±5°, based on combined ROMAP, RPC-MAG, CONSERT and Solar Array measurements.

Start Time (UTC)	End Time (UTC)
09:00:01	10:36:10
11:30:32	12:00:09
12:54:32	12:54:50
13:32:32	13:36:50
14:41:43	14:52:45

These rotation axis alignment angles are also illustrated in Fig. 2.7 while the results are given in Fig. 4.3. To resolve temporal changes, the entire interval was split up in 187 individual segments of 20 min overlapping each other by 90%. While β remained relatively stable at around $\beta=10°$, α increased from -10° to 10°, which suggests a small shift in the rotation axis over time. Based on previous work (Heinisch and Auster 2015, Heinisch et al. 2016) an error of approx. ±5° can be expected. In a final third step, the phase of the rotation around this axis was determined using a brute-force based approach. The magnetic field data was numerically rotated using the previously determined frequency while varying the rotation phase to get the best match to the corresponding RPC-MAG observations. To account for changes in magnetic filed conditions, the entire descent interval was split up into sections of 20 min, which were then independently used as input. In contrast to the quasi-static z-axis alignment, the length of the intervals had to be increased to account for the rotation period. Overall it was possible to reconstruct the full lander attitude with an accuracy of better than ±5° for most of the descent (see Table 4.1). The attitude information resulting from this analysis was converted to the SPICE format (Acton 1996) and published as ancillary information as part or the Rosetta data archives.

4.2 First Touchdown

Philae touched the surface of comet 67P on November 12, 2014 at 15:34:03.98 ± 0.10 s UTC (Biele et al. 2015). The location of the contact, later named Agilkia, was only 112 m away from the intended target. After touchdown the ADS hold-down thruster and the harpoons on the landing gear (see section 2.5.1) failed to fire. The failure of the ADS was caused by a stuck propellant gas valve. Even though this problem was detected during pre-separation checks, it was deemed unrecoverable and the separation was carried out as planed, hoping that the harpoons and ice screws would sustain enough force to anchor Philae and prevent a rebound. Earlier ground tests revealed possible problems with the nitrocellulose gas generators of the harpoons caused by long term vacuum degradation. Hence, it was decided to alter the firing sequence after touchdown to reduce the current applied to the bridge-wire detonators (Thiel et al. 2003) to prevent a premature filament failure before the nitrocellulose ignited. It was planned to automatically fire both harpoons in parallel to reduce the current and automatically retry the redundant firing system 10 s later.

The first harpoon firing command was logged at 15:34:06 UTC followed by the firing command for the redundant secondary branch at 15:34:13 UTC. No movement of the anchoring wire attached to the harpoons was detected, which indicated that the firing failed. Even though the damping mechanism dissipated a large amount of the impact energy, the ice screws alone were not enough to anchor Philae to the surface and prevent a rebound. The exact circumstances leading to this failure are still unclear, but it was possible to use ROMAP measurements as part of the post operations analysis to confirm, that the relays used to arm the system were correctly switched on during descent. Fig. 4.5 shows the By-component of the magnetic field observed by ROMAP, with RPC-MAG being used as reference. The first relay was switched on at 15:52:34 UTC causing a jump of -5 nT in the magnetic field, which was not observed on the orbiter, confirming that this event was indeed lander-based. Switching of the second relay compensated the magnetic field created by the first as intended by design, causing a secondary jump of 5 nT and the same process repeated for relays three and four. This analysis not only confirms, that the harpoons were armed correctly, but also illustrates that magnetometers can be used effectively as housekeeping tool to monitor onboard operations. Further analysis revealed the most likely cause of the harpoon failure to be a mixup between the primary and redundant power supplies, preventing current flow through the filaments.

The deceleration during touchdown caused the ROMAP boom to move downward relative to the lander due to inertia. As described in section 2.5.1 the boom is kept in its nominal position by two springs (see Fig. 2.11), allowing the boom to move when accelerated. Movement of the boom relative to the static lander bias field causes a characteristic signature in the observed magnetic field. Based on the duration of the disturbance, the corresponding contact duration was determined. It was also possible to derive the direction of boom deflection based on the sign change in the magnetic field. The influence of the boom movement on the magnetic field observations for all surface contacts is illustrated in Fig. 4.4, showing the Bx component of the ROMAP measurement during touchdown as example. A downward motion of the boom relative to the nominal position is linked to a decrease in the observed field, while an upwards motion causes an increase in the Bx component. The sharp decrease in the observed field strength during the 1st touchdown translates to a downward deflection of the boom by ~ 50 ° based on the bias field observed before boom deployment during descent.

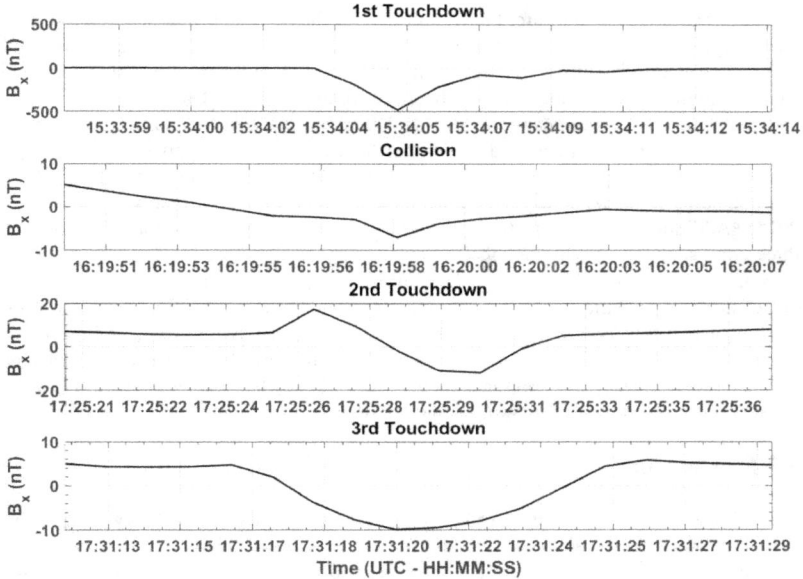

Figure 4.4: Magnetic field signature caused by the movement of the spring supported ROMAP boom relative to the lander bias field for all surface contacts, the Bx component of the ROMAP measurement is shown as an example.

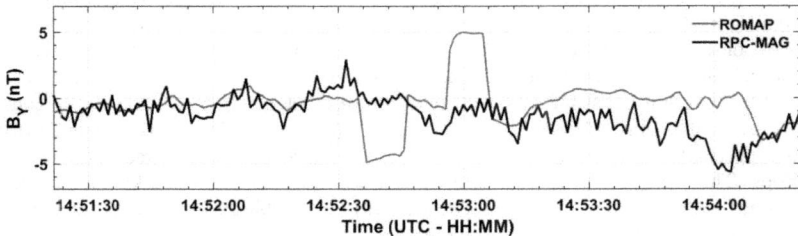

Figure 4.5: Magnetic field signature caused by the relays used to arm the harpoons, the By component of the ROMAP measurement is shown as an example with the corresponding component of the RPC-MAG observations as reference.

4.3   Rebound

After the harpoons failed to fire, Philae bounced off the surface, tilting between the individual legs while skidding across the surface, creating three distinct craters approximately 2 m wide and up to 20 cm deep (Biele et al. 2015). Based on the magnetic signature created by the movement of the ROMAP boom, the contact took at least 6 s (see Fig. 4.4) and caused at first a significant downwards deflection followed by and upwards deflection of the sensor. The internal damping mechanism dissipated approximately 21 J of the initial 49.5 J of kinetic energy, while secondary effects inside the lander (i.e. friction in the landing gear) were responsible for an additional loss of at least 5 J (Biele et al. 2015). Additional energy was dissipated during the excavation of the material from the three craters, but as only limited observational evidence about the mechanical properties is available, it is impossible to derive a meaningful estimate of the required energy. At the end of the descent Philae was rotating with a frequency of about 1.7 mHz (see Fig. 4.1), directly after liftoff the rotation rate increased to approximately 32 mHz (see Fig. 4.6 in section 4.3.1). This increase in rotation frequency must have been caused by a transfer of translational energy into rotational energy, most likely while one of the feet in contact with the surface acted as a fulcrum, allowing Philae to pivot.

4.3.1   Collision

After Philae lifted off, it started to spin up again, due to the internal flywheel. Even though the drive motor was switched off after the touchdown signal was detected by the internal electronics, the flywheel transferred some of its momentum to the lander by internal friction. This spin-up process could accurately be tracked in the spectra of the ROMAP observations, as was done before for the descent. Fig. 4.6 shows the dynamic power spectral densities for the ROMAP Bx- and Bz- components and overlaid in red the theoretically expected (based on the known moment of inertia of flywheel and lander as described by Roll and Witte (2016)) rotation frequency. The bottom panel shows the predicted and reconstructed lander rotation frequency, which reached its maximum of 77 mHz shortly before 16:20 UTC. To extract only the lander rotation frequency, a band-pass filter was used to remove all signal components from the timeseries outside the range of 30 mHz to 100 mHz (based on the spectra and a comparison with the undisturbed RPC-MAG observations). In a second step, the time between the peaks in the filtered signals was calculated to get the rotation frequency for a given time. Up to 16:20±1s UTC the rotation pattern matches the expected curve almost exactly. Afterwards, instead of continuing to spin with a relatively constant frequency (because the flywheel has spun down) as expected, the observed rotation pattern changes drastically and the rotation frequency drops. This change in rotation pattern was a clear indication that some kind of contact with the surface must have occurred, as there are no possible internal causes for such a drastic tilt. In contrast to the first (and the last two) touchdowns only a weak magnetic signature due to ROMAP boom movement was present in the magnetic field observations (see Fig. 4.4). This indicates that different than before, no significant acceleration in the z-direction of the lander was present. Therefore, this event was only classified as a collision with the comet and not as a real touchdown. The approximate geometry of this collision with the rim of the Hatmehit crater is illustrated in Fig. 4.7. It is based on the lat-

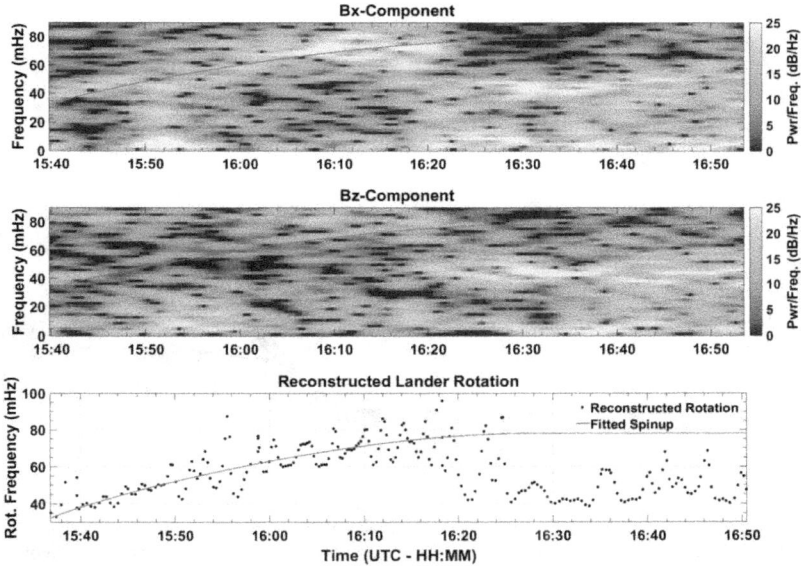

Figure 4.6: Dynamic power density spectrum of the Bx- and Bz-components of the ROMAP magnetic field observations after TD1 (15:34 UTC) with the predicted lander rotation frequency (red curve) and the reconstructed rotation frequency (bottom panel). The change in rotation pattern before and after the 16:20 UTC collision event can be seen clearly. The ROMAP By-component was omitted, as it is largely identical to the Bx-component (based on Heinisch et al. 2017a).

est OSIRIS SHAP7 (Preusker et al. 2017) digital terrain model (DTM), the reconstructed attitude and the approximate collision coordinates. The results were cross checked using the older SHAP5 V1.0 (Jorda et al. 2016) DTM to exclude errors caused by DTM artifacts. The attitude was determined from the magnetic field observations, using a similar approach to the descent. First the alignment of the rotation axis was reconstructed to be able to determine the exact phase of the rotation. To accomplish this, the same approach as previously for the descent was used. As the rotation frequency was already known, only the phase had to be varied and the results compared to the concurrently measured RPC-MAG data. This way a mean correlation coefficient between all components of $\rho >$ 0.85 was achieved. The coordinates of this collision were based on the trajectory reconstructions done by the SONC Philae flight dynamics team (Jurado et al. 2016), (Garmier et al. 2015), taking into account the exact flight time as determined from the ROMAP observations and projecting these results onto the latest OSIRIS DTM. This way the approximate collision site coordinates were determined to be [2.36; -0.38; -0.12] km using the CHEOPS reference frame. Assuming all these intermediate results to be correct, Philae struck the comet first with the +x-pointing foot with an angle of approx. 55° relative to the surface and a horizontal velocity of approx. 0.23 m/s. This attitude was again checked against the solar array illumination patterns, which confirmed these estimations.

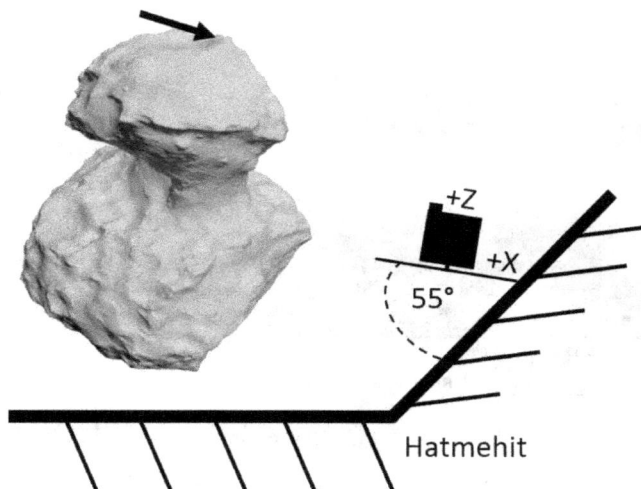

Figure 4.7: Illustration of the geometry (not to scale) of the 16:20 UTC collision event based on the attitude as reconstructed from ROMAP and RPC-MAG observations. The position of the collision on the comet is indicated by the arrow and the shape model on the left side (based on Heinisch et al. 2017a).

The collision caused Philae's axis of rotation to tilt significantly, while the rotation frequency declined to approx. 42 mHz due to the lost energy. Using the same approach as during the descent to calculate the rotation axis alignment from the magnetic field observations, the new rotation axis was tilted approx. 30° against the lander x-axis and -15° against the y-axis. This tilt in the rotation axis caused the rotation signature to become visible in the spectrum of the ROMAP z-component (see Fig. 4.6), which is not possible by a simple rotation in the x-y plane, as during descent and before the collision. In addition to the rotation Philae also nutated with a period of 453 s. Because the internal flywheel had completely spun down by 16:20 UTC the dynamic lander attitude was now only governed by its moments of inertia. As this rotation pattern was much more complicated than before the collision, only the attitude directly after the collision and before the second touchdown was determined. The detailed attitude results for the period before the collision will be published in the PSA archive of ESA and the PDS archive of NASA as part of the ancillary mission information (Vazquez-Garcia et al. 2015).

4.3.2   Second Touchdown

At 17:25:35±1s UTC another boom movement was detected in the ROMAP magnetic field observations (see Fig. 4.4). Together with a complete change in the dynamic lander

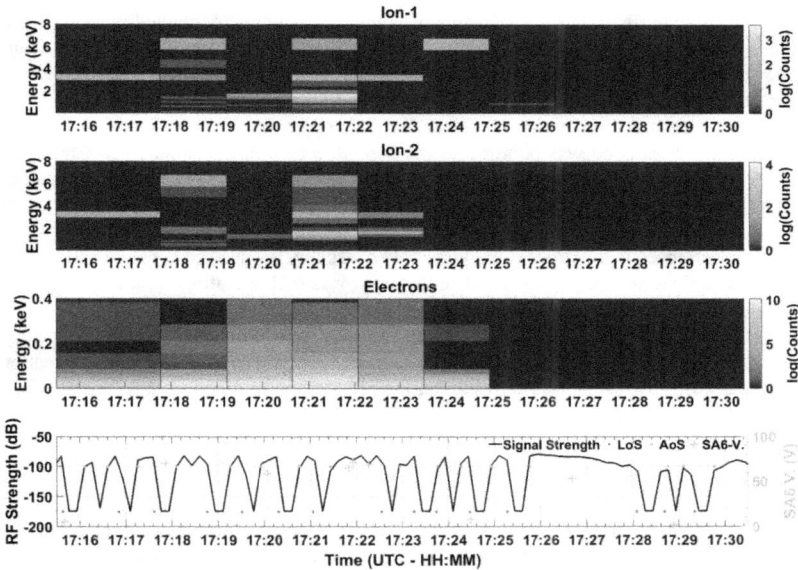

Figure 4.8: ROMAP SPM particle counts for the Faraday cup and the two ion channels showing the sudden drop-out after 17:25 UTC observed on November 12. As comparison the RF-Link signal strength with lander signal acquisition (AoS) and loss off signal (LoS) and the voltage of solar array 6 (SA6) on the lid are displayed (based on Heinisch et al. 2017a.

attitude after this event, it was a clear indication of another TD, during which Philae lost most of its remaining energy and momentum. Little is known about the details of this second touchdown as there are no detailed images of the touchdown site. Based on the flight time, surface topography and the known final landing site, this contact most likely happened in the direction of [-0.152; -0.952; 0.266] km relative to the final site in an area roughly at [2.445 -0.072 -0.343] km in CHEOPS coordinates with a horizontal velocity of approx. 0.16 m/s. Fig. 4.8 displays the particle counts for the ROMAP plasma monitor SPM (see Fig. 2.7 for the position of the SPM sensor) ion and electron detectors in addition to RF-Link and solar array housekeeping data. The solar wind particle (ions and electrons) counts drop out immediately after this touchdown incident and remained zero for the rest of the SPM observation period. As ROMAP housekeeping measurements remained nominal and the magnetometer continued to operate as expected, mechanical or electrical destruction of the sensor can be excluded. Especially the high-voltage needed to operate SPM is known to cause detectable arcing if the sensor is compromised in any way.

The top solar array (SA6) was not illuminated at 17:24:30 UTC and moved into Sunlight until 17:26:41 UTC. Philae lost the RF-Link with the Rosetta orbiter at 17:25:16 UTC (LoS) and reacquired the signal at 17:25:46 UTC (AoS). Afterwards the link remained stable until 17:28:05 UTC. This behaviour can only be explained by shadowing caused by terrain or by Philae pointing away from the Sun and the orbiter. Shadowing by surrounding terrain is rather unlikely considering the local terrain. Based on the earlier behaviour after collision it is much more likely, that off-pointing was the cause for the link break and illumination changes. Because of geometric constraints, this would imply that the balcony was pointing in the general direction of the surface. The most likely scenario based on these observations is that the balcony tipped towards the comet surface shortly before the time of contact (most likely due to the rotation pattern) and the boom touched the upper surface layers scooping up cometary material that blocked the detector entrances in the process. A dust cloud caused by the surface contact itself would not be sufficient to cover up the SPM sensor, as it was still operating nominally after the much more powerful first touchdown, which is known to have created a sizable dust cloud (Biele et al. 2015). The particle drop-out can also not be caused solely by SPM pointing (i.e. continuously pointing at the surface) because no particles were detected even after the measured SA6 voltage shows illumination of the panels, which requires the lid (and SPM) to face towards the Sun.

4.3.3   Final Approach and Touchdown

After the second touchdown, the rotation of the lander was slowed down significantly to only ~ 1.0 mHz. The axis of rotation was close to the lander y-axis. Approximately 40 s before the beginning of the final surface contact at 17:31:16 ± 1 s the rotation stopped and the attitude remained stable. As only a limited set of magnetic field observations was available, the attitude was not solely determined based on the magnetic field. Additionally the RF signal strength (Dudal and Loisel 2016) and solar array illumination was also considered to constrain the attitude. To achieve the observed RF signal strength of -91 dB, the panel antennas of the lander must have been pointing towards the Rosetta orbiter with a maximum deviation angle of 35°. Only solar panel SA2 was generating any meaningful power (~ 1.1 W) hence, this array was pointing in the general direction of the Sun and partly shadowed and illuminated at an oblique angle. Theses results are illustrated in Fig. 4.9, showing a model of Philae with the appropriate attitude overlaid over the image of the final landing site (adapted from Sierks 2016), the Rosetta and Sun direction vectors are depicted as grey and yellow lines respectively. Based on this attitude, the balcony was roughly parallel to the local surface, with the lander y-axis tilted slightly relative to the direction of flight. The final attitude after the third touchdown was determined from combined RPC-MAG and ROMAP magnetic field observations (Heinisch et al. 2016) with an error below ±5°. This attitude information was successfully used to reestablish contact with Philae in June 2015 (Ulamec et al. 2017). The translation of the lander was most likely stopped by a cliff-like structure and Philae then touched down with its feet on the comet surface, accelerating the ROMAP boom one last time during the third touchdown. OSIRIS-NAC images taken on September 2, 2016 clearly show Philae at the final landing site Sierks (2016) and made it possible to accurately determine the landing site coordinates. Fig. 4.10 shows one of the NAC images overlaid with two rendered models of Philae illustrating the attitude directly before the third touchdown (left) and the

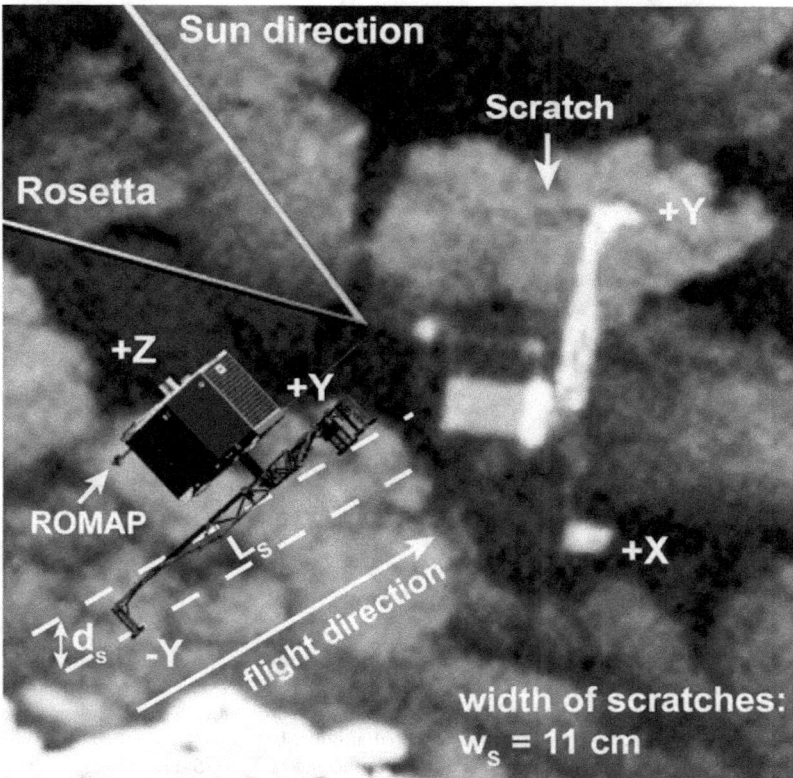

Figure 4.9: OSIRIS NAC image of Philae at the final landing site overlaid with a model of Philae illustrating the reconstructed attitude for the last ~ 40 s before TD3. The Rosetta and Sun direction vectors are depicted as grey and yellow lines respectively (image adapted from Sierks 2016).

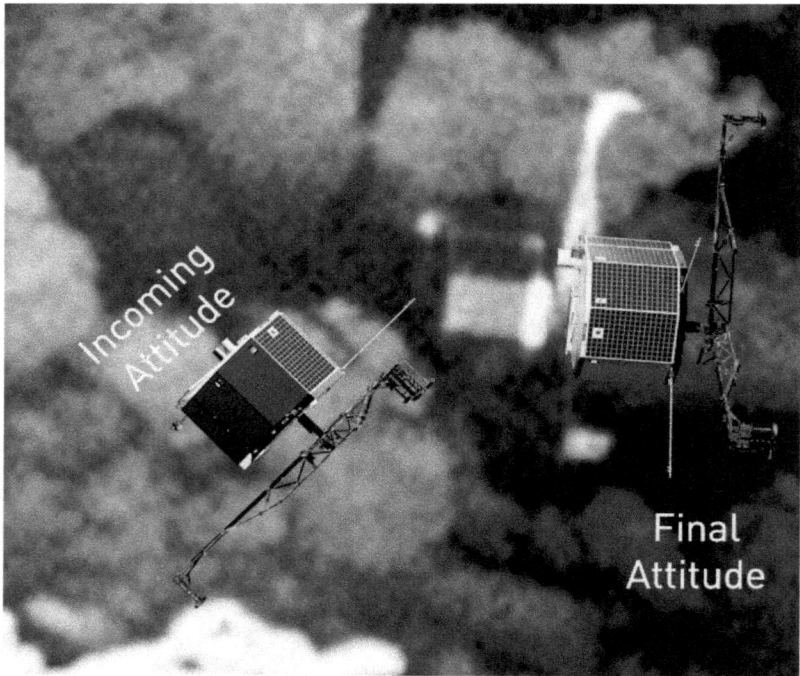

Figure 4.10: OSIRIS NAC image of Philae at the final landing site overlaid with two models of Philae illustrating the reconstructed ROMAP attitude before (left) and after TD3 (right - position shifted for visibility) (image adapted from Sierks 2016).

final attitude (right - position shifted for better visibility) the attitude changes commanded at the end of the FSS (translation and rotation, Biele et al. 2015) were taken into account to be able to relate the OSIRIS image to the attitude information. These images not only confirm the accuracy of the reconstructed attitude, but also exclude the possibility of any significant attitude change of Philae after the end of the FSS due to outgassing or sublimation.

5   67P/Churyumov-Gerasimenko Surface Magnetization

The magnetization of comets is not only important because magnetic fields may have played a role in the formation of such objects (Nübold and Glassmeier 1999, Fu and Weiss 2012), but it also allows to constrain the strength of the background magnetic field in the early solar system (Wang et al. 2017). For that reason, in addition to the magnetometers, the atomic force microscope MIDAS (Riedler et al. 2007) onboard Rosetta was equipped with magnetic tips to perform magnetic force microscopy. Because of technical reasons, these tips were not used and the magnetization of the cometary grains could only be determined based on magnetometer measurements. Using concurrent observations from the Rosetta orbiter magnetometer RPC-MAG and ROMAP, Auster et al. (2015) derived an upper limit for the specific magnetic moment of $<3.1 \times 10^{-5}$ Am^2/kg for meter-sized homogeneous boulders. The upper bound for any influence caused by local surface magnetization on the observed magnetic field was chosen to be 2 nT. As the final location of Philae was still unknown, Auster et al. (2015) used an estimated trajectory to reconstruct the distance between the lander and the cometary surface.

After the final landing site of Philae was identified using OSIRIS camera images (Sierks 2016) it was possible to further constrain Philae's trajectory, as explained in chapter 4. With a more detailed knowledge of the magnetic field (see chapter 3) around the comet, it was also possible to lower the upper bound for any influence caused by local surface magnetization. Hence, the magnetization of 67P was revisited (Heinisch et al. 2018b) using the newer results not available for the initial study by Auster et al. (2015).

After the initial touchdown at 15:34:04 UTC (TD1) on November 12, 2014, the lander rebounded, colliding with the surface at 16:20:00 UTC (COL), touching down again at 17:25:35 (TD2) and coming to a final rest at 17:31:16 UTC (TD3). An analysis of the ROMAP observations also revealed that the magnetometer boom assembly came in contact with the surface material during the second touchdown. As a result parts of the sensor head were covered with enough surface dust to prevent the ROMAP plasma monitor from detecting any solar wind particles (as explained in section 4.3.2). The diameter of the ROMAP instrument is approximately 10 cm with the fluxgate magnetometer located in the middle and the plasma sensors placed on the outer part of the sensor head. The entrances to the particle detectors on the top and sides of the instrument have a width of approximately 3 cm. Based on the fact that the dust did not fall off during the remaining flight or alter the deployment angle of the boom, the dust aggregates had to be small relative to the size of the sensor. Accordingly, the dust grains blocking the entrances of the sensor had to be smaller than ~ 3 cm, which is the size of the sensor openings. This is also consistent with particle size distributions estimated by Blum et al. (2017).

Auster et al. (2015) assumed a conservative estimate of the distance between the magnetometer and the surface between 1.2 m and 0.4 m depending on location. While the distances of 1.2 m for the 1st touchdown and 0.4 m for the collision and final touchdown still apply, a much smaller value of 0.05 m (half the sensor head diameter) can be used for the second touchdown as the boom hit the surface. Hence, it is possible to resolve much smaller grains in the range of the sensor - surface distance.

The magnetic field observations during descent and landing were dominated by magnetic low frequency waves, studied in detail by Richter et al. (2016) and described in section 3.2. These singing comet waves (Richter et al. 2015) with frequencies of 5 mHz to 50 mHz are created by the interaction between the outgassing comet and the solar wind (Glassmeier 2017). During the descent and rebound, they are the main cause of the variations in the magnetic field. Hence, analyzing these waves is a prerequisite when using the magnetic observations to gain insight into the magnetization of the nucleus. The magnetic field measurements also revealed that no magnetic boundary layer exists above the cometary surface, which makes it possible to use the observations of the two RPC-MAG magnetometers in orbit as direct reference for the surface measurements as explained in section 3.2. Using the described results, the magnetization of comet 67P is revisited.

5.1   Observations

The ROMAP magnetometer was operating during the entire descent and rebound phase with a constant sampling rate of 1 Hz (Auster et al. 2015). While the RPC-MAG outboard sensor was sampling at a rate of 20 Hz, the inboard sensor measurements are available at a rate of 1 Hz. The RPC-MAG data was processed as described by Richter et al. (2016). To use the RPC-MAG outboard-sensor observations as reference, the measurements were downsampled to 1 Hz to match ROMAP and RPC-MAG inboard observations. While the RPC-MAG inboard sensor was primarily intended for calibration purposes and as possible backup, the measurements can be used for direct scientific analysis as well. As the RPC-MAG sensors are mounted 15 cm apart on a boom placing them only 1.5 m away from the main Rosetta body, spacecraft interference is still present in the observed magnetic field as explained by Richter et al. (2015). Because of the spatial separation between the IB and OB sensors and the inhomogeneous spacecraft field, the influence on the two sensors is different.

5.1.1   Sensor Offsets and Lander Residual Field

In this section, the sensor offsets for the different parts of the descent and rebound are determined based on the lander rotation pattern. As Philae was rotating during the entire rebound with a known spin rate (see section 4.3), an in-flight calibration of the sensor-offsets and lander residual field was possible. Residual field effects can not always be

Table 5.1: ROMAP offsets in instrument coordinates determined immediately before and after each contact.

	Bx (nT)	By (nT)	Bz (nT)
Before TD1	-0.7	1.5	3.9
After TD1	-0.7	1.5	3.9
Before Col	-0.7	1.5	1.0
After Col	-0.7	1.5	1.0
Before TD2	-3.1	2.5	3.6
After TD2	-3.1	2.5	3.6

separated from sensor offsets. Hence, in the following "sensor offsets" is used for the reason of simplicity for the sum of the lander residual field and sensor offset.

With exact knowledge of the offsets, it is possible to determine the calibrated magnitude of the magnetic field, which is independent of the magnetic coordinate system. Hence, by use of the magnitude, it is possible to relate the ROMAP observations to the RPC-MAG orbiter measurements, without introducing errors due to attitude uncertainties. Auster et al. (2015) initially assumed constant offsets for all contacts, as the sensor temperature only changed by about ~10°C during the rebound. In general, constant sensor offsets do not interfere with the determination of the surface magnetization, but can introduce an additional bias when comparing orbiter and lander measurements. To further reduce uncertainties in the ROMAP observations, the offsets were re-determined individually for the periods before and after each surface contact.

The offsets of the Bx- and By-components (in the reference frame of the sensor) remained stable for the first part of the rebound. An automatic change from descent to surface operations after the 1st touchdown caused a change of 1.5 nT in the Bz-component before the collision, due to a change in the lander bias field. This change was limited to one component of the magnetic field, due to the layout of the relevant electronics. After the collision, the switch off of the CIVA/ROLIS (Bibring et al. 2007a) camera system at 17:21 UTC caused an additional bias field change in all three components. The derived offsets are summarized in Table 5.1.

5.1.2   Contact Velocities

The velocities for descent and ascent were used to relate the observed magnetic field directly to the height above the surface. Initially Philae had a descent velocity of -1 m/s relative to the comet (Biele et al. 2015, Roll and Witte 2016) and a climb rate of ~0.33 m/s after the 1st touchdown. For the inbound and outbound velocities for the remaining trajectory the flight reconstruction of section 4 was used. This reconstruction is based on a free ballistic flight, with known contact times (see section 4.3), taking into account OSIRIS orbiter camera images of Philae in flight and using the latest SHAP7 digital terrain model (Preusker et al. 2017). The velocities used below are mean values based on the evaluation of several possible trajectories. As the velocity is low relative to the magnetometer sampling rate, errors in the velocity have no significant influence on a possible dipole signature in the magnetic field observations. The descent velocity for the collision that followed after the 1st touchdown was approximately -0.33 m/s and the lander climbed again with a velocity of ~0.22 m/s which was also the descent velocity for the following second touchdown. Afterwards Philae traveled for an additional ~ 284 s which equates to about ~10 m before coming to a final stop at Abydos. During that last part of the flight Philae reached a maximum height of ~1.5 m above the surface assuming a free ballistic trajectory.

5.1.3 Magnetic Field Observations

Any magnetic field created by a magnetized cometary surface should only influence the Philae ROMAP observations and not the RPC-MAG measurements in orbit. Hence, the undisturbed RPC-MAG observations in orbit can be used as reference. Fig. 5.1 shows the magnitude of the magnetic field observed by ROMAP and the RPC-MAG inboard (RPC-IB) and outboard (RPC-OB) magnetometers during descent/ascent of Philae (based on Heinisch et al. 2018b). As Auster et al. (2015) already excluded any larger scale magnetization, the measurements are shown up to a maximum height of 10 m above the surface. The observed field for descent and ascent is displayed for all contacts, except for the flight after the second touchdown towards the final third touchdown as Philae never exceeded a height of ~1.5 m. Hence, for this last part, a 200 s interval of magnetic field observations is shown, as the exact height above the surface is still uncertain. All three sensors show almost identical signals, with a mean correlation coefficient of $\rho \sim 0.78$ between RPC-MAG and ROMAP. As both magnetometers measure almost the same magnetic field, even though RPC is in orbit ~ 20 km away from Philae, there are no significant near-surface effects influencing the magnetic field. Small differences in the magnitude observed in orbit compared to the surface measurements are caused by effects of the near surface coma or remaining measurement uncertainties.

To further analyze the signals, the difference between the magnitudes of ROMAP and the two RPC-MAG sensors was calculated. This difference signal is shown in Fig. 5.2 for ROMAP and the RPC-MAG outboard sensor (RPC-OB) as illustration. While the descent observations are shown in blue, the ascent is shown in red. No distance-dependent dipole signature is observable in the difference between the ROMAP and either RPC-MAG inboard or outboard sensor measurements. The standard deviation averaged between the difference between the ROMAP and RPC-MAG IB and OB observations was used to define an upper limit for any possible contribution of surface magnetization to the total observed magnetic field.

Auster et al. (2015) used the mean value of the averaged absolute values between 10 m and touchdown for descent and ascent to estimate an upper limit of 2 nT for any contribution to the observed field from surface magnetization. As the exact height of Philae during the fight between the second and third touchdown is still unknown, this approach can only be used for the 1st touchdown and collision. For this study the standard deviation of the difference between ROMAP and RPC-MAG was chosen as an estimate for an upper limit. It is not only independent of the specific height above the surface, but also invariant to remaining offsets caused by spacecraft bias fields, which are constant during

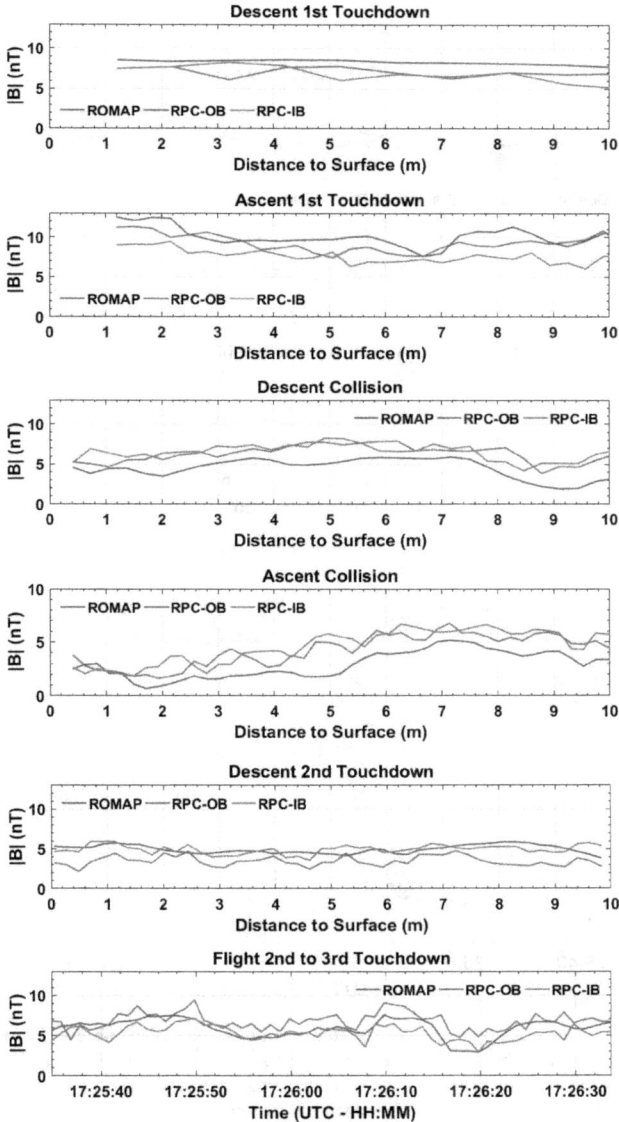

Figure 5.1: Magnitude of the magnetic field observed by ROMAP (red), RPC-MAG OB (blue) and RPC-MAG IB (green) during the different surface contacts. As the exact height over the surface after the second touchdown is unknown, the magnetic field is shown versus time instead of distance to the surface (based on Heinisch et al. 2018b).

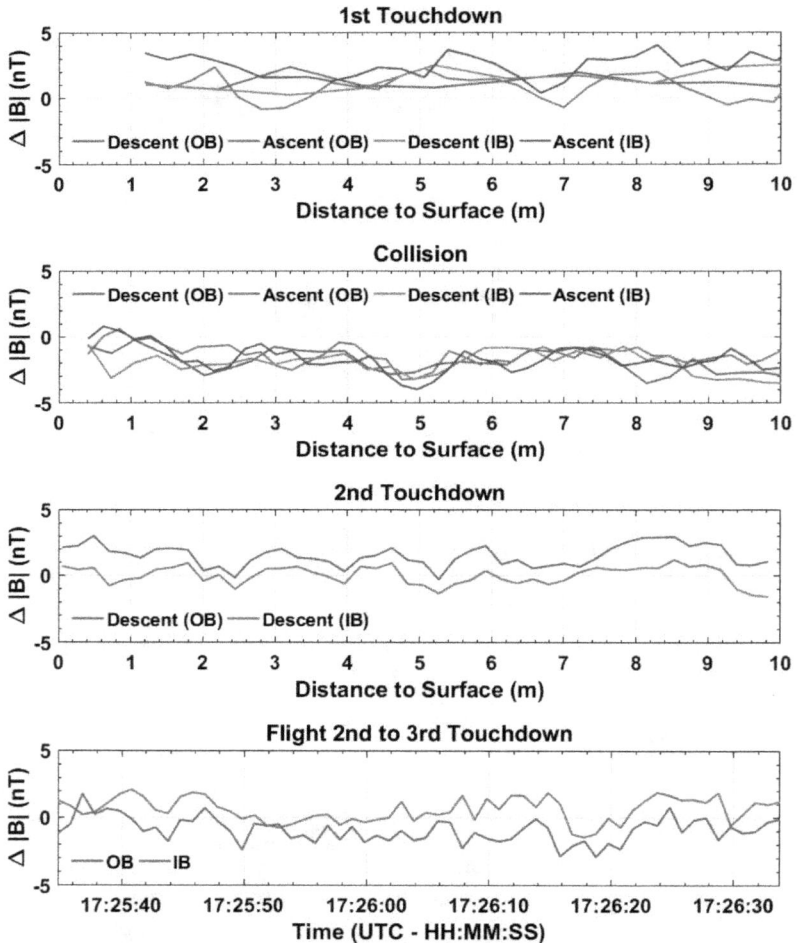

Figure 5.2: Differences between the magnitudes of the magnetic field observed by ROMAP and the RPC-MAG inboard (IB) and outboard (OB) sensors during descent and ascent of Philae. As the exact height over the surface after the second touchdown is unknown, the magnetic field for the climb after the second touchdown is shown versus time instead of distance to the surface (based on Heinisch et al. (2018b)).

Table 5.2: Standard deviation between the difference of the RPC-MAG and ROMAP observations for the final 10 m before contact.

	Descent	Ascent
Touchdown 1	0.68 nT	0.89 nT
Collision	0.86 nT	0.88 nT
Touchdown 2	0.76 nT	0.91 nT

the entire interval of interest. A height dependent dipole-like drop off in the magnetic field potentially caused by surface magnetization would cause additional variance in the observed field, increasing the standard deviation and hence the later derived limit for the magnetization. Additionally the standard deviation also accounts for the variability of the magnetic field between Philae and Rosetta, which could mask the dipole signature caused by small scale magnetization of the surface. Table 5.2 shows the standard deviation for the difference between RPC-MAG (averaged over IB and OB) and ROMAP during the ascent and descent respectively. Based on these results a rounded upper limit of 0.9 nT was chosen for any influence caused by local surface magnetization on the observed magnetic field.

5.2 Magnetization

To determine the original small scale magnetization of comet 67P, it was necessary to assume, that the measured surface grains were relatively pristine, preserving their initial magnetic properties. Processing of the grains, primarily due to repeated ejection and fallback could alter the magnetic properties. While the area around the initial 1st touchdown is covered by a heavily processed fallback-dust layer (Biele et al. 2015), the areas surrounding the collision and the second touchdown sites are more consolidated (Schröder et al. 2017, Keller et al. 2015, Thomas et al. 2015). Therefore an underestimation of the magnetization of the surface material due to coverage with processed dust can be excluded (see also section 6.3). Using the model proposed by Auster et al. (2015) based on randomly oriented, uniformly magnetized cubes with a magnetic dipole moment $\vec{m}$, an estimate for the magnetic field B caused by a single dipole at the center of the cube can be derived:

$$\vec{B}(\vec{r}) = \frac{\mu_0}{4\pi} \frac{3\vec{r}(\vec{m}\vec{r}) - \vec{m}\vec{r}^2}{|\vec{r}|^5}, \tag{5.1}$$

where $\vec{r}$ is the spatial vector from the dipole center to the magnetometer and μ_0 is the vacuum permeability. Using $\vec{m} = \vec{M} \cdot D^3$ where D is the scale of the cubes or the spatial resolution and $\vec{M}$ is the magnetization, this becomes:

$$\vec{B}(\vec{r}) = \frac{\mu_0}{4\pi} \frac{3\vec{r}(\vec{M}D^3\vec{r}) - \vec{M}D^3\vec{r}^2}{|\vec{r}|^5}. \tag{5.2}$$

Based on the modeling performed by Auster et al. (2015), a single dipole in the second Gaussian orientation (dominating horizontal magnetic moment) can be used to derive a conservative estimate for the scalar magnetization M:

$$M \leq \frac{4\pi B}{\mu_0} \left(\frac{r}{D} + \frac{1}{2} \right)^3 \tag{5.3}$$

with $B = 0.9$ nT being the upper limit for the magnetic field and the scalar $r = 0.05$ m being the closest distance between the material and the sensor (see above). The spatial resolution D of the resolvable particles was chosen to be $D = 0.05$ m, based on the measurement and sensor geometry. The spatial scale of the derived magnetization is given by the ratio between r and D. With an r/D ratio of 1, this results in a maximum magnetization of $\sim 3 \cdot 10^{-2}$ A/m. Using a mean mass density of 533 kg/m^3 (Pätzold et al. 2016) this translates to an upper limit for the specific magnetic moment of $\sim 5 \cdot 10^{-5}$ Am^2/kg for grains in the range of ~ 5 cm. The reduction of the spatial resolution to particles with a size of ~ 5 cm makes it possible to resolve structures in the sub-decimeter size range, which is characteristic for cometary agglomerates (Blum et al. 2017) and as described earlier, for the particles the sensor came in contact with. This is a significant improvement in resolution compared to the more theoretical spatial resolution of D = 1 m with a corresponding r/D ratio of 0.4 assumed by Auster et al. (2015) based on the larger sensor - material distance of 0.4 m. Even though the spatial resolution improved by approximately two orders of magnitude, the resulting maximum specific magnetic moment is still comparable with the previous estimate of $\sim 3 \cdot 10^{-5}$ Am^2/kg by Auster et al. (2015). Assuming larger homogeneously magnetized areas on the surface in the range of D ~ 1 m as used by Auster et al. (2015), the specific magnetic moment derived based on this analysis decreases by an order of magnitude to $\sim 2 \cdot 10^{-6}$ Am^2/kg. Hence, this is a significant improvement over the results of Auster et al. (2015), as it was possible to show, that the nucleus is remarkably unmagnetized down to the scale of individual agglomerates and not just for boulders in the range of ~ 1 m.

5.3   Magnetic Field in the Solar Nebula

Results from the Stardust mission suggest that in general magnetically susceptible material is present on comets (Ogliore et al. 2010, Berger et al. 2011). If comets like 67P are collisional remnants as conjectured by Morbidelli and Rickman (2015), no conclusions can be drawn with regard to the magnetic environment during creation of the comet. As collisions would have caused significant reprocessing of the material, the observed magnetization would provide no information about the magnetic conditions during the initial growth of the nucleus.

If comets are instead primordial rubble piles which experienced negligible thermal or aqueous reprocessing after formation as described by Davidsson et al. (2016) and Blum (2018), the lack of detectable magnetization might provide insight into the magnetic conditions inside the solar nebula. As discussed in chapter 6, the primordial rubble pile model is also consistent with the compressive strength derived based on the Philae surface contacts. Two major mechanisms are known to lead to the creation of magnetized particles, accretional detrital remanent magnetization (ADRM) as explained by Fu and Weiss (2012) and accretional attractive remanent magnetization (AARM) as described by Nübold and Glassmeier (2000). Simulations have shown, that even in the absence of an external field, AARM can cause the growth of aggregates up to several mm in size by the accretion of further magnetic material by single magnetized seed grains (Nübold and Glassmeier 2000). ADRM can create larger coherently magnetized regions (up to decimeters in size) and is based on the alignment of magnetized particles during accretion due to the presence of an external magnetic field and can thus play a role either as a standalone effect or coupled with AARM.

Based on the observed magnetization, no magnetized regions down to scales of ~ 5 cm have been detected, which gives strong evidence for the absence of ADRM. Assuming comets are largely unaltered primordial rubble piles (Davidsson et al. 2016, Blum 2018), the most likely explanation is the absence of a strong background magnetic field during accretion. This puts a threshold on the magnitude of the magnetic field in the solar nebular during the formation of the aggregates the nucleus is made up of. To acquire ADRM, the aligning force of the background magnetic field must overcome other randomizing forces during accretion like gas drag, Brownian motion or collisional effects (Fu and Weiss 2012). Modelling of the forces relevant for ADRM performed by Fu et al. (2018) and Biersteker et al. (2018) showed, that during the approximate time of accretion, at a heliocentric distance of ~ 20 AU, the magnetic field must have been below ~ 4 μT. Otherwise the magnetic torque from the background field would have overcome randomizing effects, aligning grains in the range of ~ 5 cm and causing a significantly higher magnetic moment. This constraint not only depends on the observed magnetization, but also on the achieved spatial resolution. With a resolution of ~ 1 m, the results of Auster et al. (2015) could only be used to determine the presence of ADRM for larger boulders, above the typical decimeter size range, at which ADRM is effective (Fu and Weiss 2012). With the aggregate-level resolution achieved as part of this work, the absence of ADRM was confirmed even on the more realistic scale of individual grains, which is necessary to derive a constraint for the background magnetic field at the time of accretion.

6 Surface Compressive Strength

One of the primary objectives of the lander mission was to measure the surface properties of 67P, as they can provide insights into how comets formed and what they are made of. To measure the mechanical properties, primarily the tensile and compressive strength, Philae was equipped with the MUPUS hammer system in conjunction with the seismographic SESAME-CASSE instrument. Because of the unexpected circumstances of the landing, these instruments could not be used as intended. In this chapter, the rebound trajectory is combined with the reconstructed attitude, to estimate the energy balance for the individual surface contacts. The energy loss during these contacts was then used to derive an estimate for the compressive strength of the local surface material.

6.1 Comet Formation Scenarios

In one possible scenario comets have formed in the young Solar System by the gentle gravitational collapse of dust clouds, typically consisting of sub-decimeter-sized aggregates (Johansen et al. 2007). The nature of this process means that the nucleus consists mostly of intact dust aggregates (Blum et al. 2017), which have survived the comet formation, owing to small impact velocities of less than 1 m/s during collapse and have thus experienced relatively little compression (Wahlberg Jansson and Johansen 2014, Wahlberg Jansson et al. 2017). In the other possible scenario, comets are believed to have formed by hierarchical agglomeration, which is based on growth by high velocity impacts (up to ∼ 50 m/s) onto the cometesimal to form the cometary nucleus. Compared to a gentle gravitational collapse, this leads to an increased mechanical strength and volume filling factor due to the higher compaction of the material (Blum 2018). Hence, in situ measurements of the mechanical properties like the compressive strength allows to test these formation hypotheses. The most likely primordial nature of 67P (Davidsson et al. 2016) makes it an ideal candidate for these types of studies.

Several different estimates for the compressive strength σ have been published. Spohn et al. (2015) derived a uniaxial lower limit of 2 MPa based on the results of a spot-measurement with the Philae MUPUS hammer experiment. Biele et al. (2015) and Roll et al. (2016) estimated a surface compressive strength in the order of 1 kPa based on an analysis of the mechanics of Philae's first touchdown. By analyzing the collapse of cliff overhangs observed from orbit, Groussin et al. (2015) inferred a compressive strength of between 30 Pa and 150 Pa.

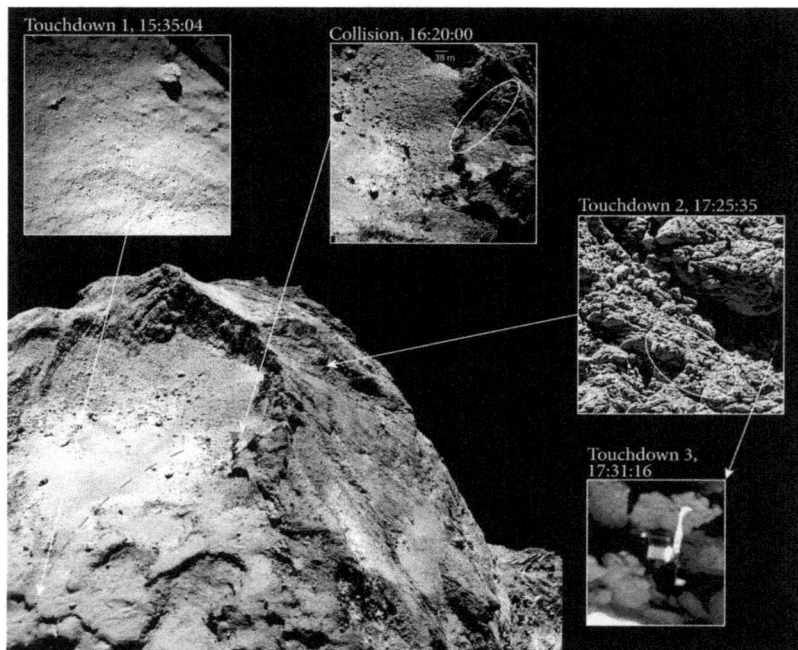

Figure 6.1: Overview of the four contact sites on the upper lobe of comet 67P. All contacts took place during Philae's descent on November 12, 2014. The zoomed-in inlays show the differences in morphology between the four contacts, the possible areas of contact are circled in white. The flight direction is illustrated by the red arcs (based on Heinisch et al. 2018b).

In the following chapter an upper limit for the surface compressive strength is derived by using the Philae lander as an impact probe (based on Heinisch et al. 2018a). The compressive strength is determined based on the energy balance and geometry of Philae's contacts. While the lander initially touched down in an area covered by fallback dust, the other contacts took place in more consolidated areas without significant fallback. Deriving the compressive strength based on the energy balance requires reasonable knowledge of the contact duration and corresponding geometry. Reconstructions of the 1st touchdown (Biele et al. 2015, Roll and Witte 2016, Roll et al. 2016) showed, that Philae pivoted several times between the individual landing feet. As the exact sequence of these events can only be estimated based on simulations, it is impossible to derive the contact geometry and corresponding contact duration. Additionally, in contrast to the subsequent surface contacts, the landing gear damper system was engaged during the initial touchdown, actively dissipating some of the kinetic energy. Hence, in this study the 1st touchdown was not revisited. The energy balance and compressive strength was only derived for the collision, second and third touchdown.

To determine the total kinetic energy balance (translational and rotational energy) for each of the surface contacts, it is necessary to know the incoming and outgoing velocities and the rotation rate of the lander during flight. While the velocities can be derived from the corresponding trajectories, the rotational energy was determined based on the rotation frequencies derived in chapter 4. In the following section, the locations of the individual touchdowns are discussed and used to derive the required velocities based on assumed ballistic flight parabolas, taking into account previous reconstructions by Auster et al. (2015), Biele et al. (2015) and the analysis from section 4.3. In a next step, attitude information gained from two-point magnetic field observations by the ROMAP (Auster et al. 2007) and RPC-MAG (Glassmeier et al. 2007b, Richter et al. 2016) magnetometers as described in section 4.3 is used in conjunction with a digital terrain model of 67P (Preusker et al. 2017) to determine the geometry of the contacts. This is necessary to calculate the contact pressure (force per contact area). The compressive strength is finally calculated based on the total energy balance, contact geometry and duration of the surface contacts.

Next, attitude information for the last part of the trajectory up until the third touchdown is combined with the scratches visible in the images of the Lander at Abydos. As these features were most likely created by Philae, they can be used to constrain the surface compressive strength as well. Finally, the implications of these results are discussed.

6.2 Philae as Impact Probe

As the first and final landing sites were photographed from orbit (see Fig. 6.1), the respective positions were determined based on these images. For the first touchdown at Agilkia (see section 4.2), the Cartesian (x, y, z) coordinates $(2.12, -0.96, 0.50)$ km given by Biele et al. (2015) were used, while for Abydos (final location of Philae) the estimation $(2.45, -0.07, -0.34)$ km derived in section 4 was used. All coordinates are given in the 67P body-fixed CHEOPS coordinate-system (SPICE: 67P/C-G_CK). The NASA NAIF SPICE system (Acton 1996) was used to facilitate the coordinate calculations.

The trajectory between the 1st touchdown and the collision was constrained by the flight time, surface structure and an image of Philae and its shadow in flight, taken approximately ten minutes (15:45:02 UTC) after the 1st touchdown (see Fig. 6.2 a). Based on this image, Philae's approximate position at 15:45:02 UTC is $(2.27, -0.88, 0.36)$ km. By combining this information with the coordinates of the first touchdown, the direction of flight was determined (see Fig. 6.2). The flight time $t_1 = (2756 \pm 3)$ s was derived based on the time between the initial touchdown and the collision. Assuming a free ballistic trajectory (similar to Auster et al. 2015) with a mean local gravitational acceleration of $g \approx 1.6 \cdot 10^{-4}$ m/s^2 (based on Biele et al. 2015), a possible area (approximately 120 m x 90 m) for the collision was identified using the latest SHAP7 digital terrain model (Preusker et al. 2017). This area was selected based on the flight direction taking into account the preliminary collision coordinates discussed in section 4.3.1, the flight time and uncertainties in the digital terrain model. As the aim of this analysis is to estimate the velocities and kinetic energy for the rebound, it is not necessary to determine the exact contact locations or definitive trajectories. To simplify computation and to be independent of a specific version of the terrain model, the area of interest was approximated by three polygons placed parallel to the local surface to reflect the different slopes of the Hatmehit crater rim.

Possible locations for the second touchdown are constrained by the estimated position of the collision (taking uncertainties in the location into account) and the known final landing site. A cliff-like structure visible in the upper right hand corner of Fig. 6.2 further restricts the trajectory, as Philae would not have been able to clear the top of this cliff (relative height ~ 60 m) on a ballistic flight coming from the collision site. Hence, the site of the second touchdown must be between this cliff and the location of the collision. This yields an estimated flight direction relative to the third touchdown given by [0.15, -0.95, 0.27] km as already explained in section 4.3.2. Based on the flight time between the second and third touchdown of only $t_3 = (215 \pm 3)$ s, the locations of the two touchdowns must be close together. A rough initial estimate for the horizontal velocity between the collision and second touchdown can be derived from the corresponding flight time $t_2 = (3926 \pm 3)$ s, preliminarily assuming the locations of the second and third touchdown to be identical. Combining this estimated velocity with the flight directions, results in the area highlighted in red in Fig. 6.2 b). The center of this region is approximately at [2.45, -0.07, -0.34] km. Even though the available information is not sufficient to determine a definitive touchdown site, the derived area is sufficient for the purpose of this study.

Using the derived touchdown positions, the velocities and corresponding kinetic energies for each possible part of the trajectory were calculated. Combined with the rotation rates determined from magnetic field observations, the total energy balance of the rebound was derived. To facilitate this, the possible areas for the individual touchdowns were discretized with a spatial resolution of 1 m in each direction and the trajectories for each of these points were calculated. To determine the horizontal and vertical velocity components during the ballistic flights, a simple horizontal motion with constant velocity v_h and a vertical motion upwards with a constant initial velocity, combined with a free fall were used. Because of the difference in relative height (Δh), the outgoing vertical velocity (v_{vO}) after contact can differ from the incoming vertical velocity (v_{vI}) of the following contact. The horizontal and vertical components of the velocity were calculated using:

$$v_h = \frac{d}{t_f}, \tag{6.1}$$

$$v_{vO} = \frac{gt_f}{2} + \frac{\Delta h}{t_f}, \tag{6.2}$$

where d denotes the horizontal distance and Δh the height difference between the touchdowns. The incoming vertical velocity is calculated based on the outgoing velocity of the previous surface contact:

$$v_{vI} = v_{vO} - gt_f. \tag{6.3}$$

The flight time t_f was determined based on the time of the individual contacts to be $t_1 = (2756 \pm 3)$ s for the flight between the 1st touchdown and the collision, $t_2 = (3926 \pm 3)$ s for the flight between the collision and second touchdown and $t_3 = (215 \pm 3)$ s for the

Figure 6.2: Panel a) shows the possible area for the collision event overlaid on the digital terrain model (SHAP7) and the estimated trajectory (constrained by red lines, direction indicated by white arrow) from the 1st touchdown including the position of Philae in flight as seen on OSIRIS camera images. Panel b) shows the possible area for the second touchdown overlaid on an OSIRIS image. The incoming flight direction is indicated by a red arrow. Philae is visible to the right at its final locations (circled) (based on Heinisch et al. 2018b).

last part of free flight after the second touchdown. The rotational energy was calculated using:

$$E_{rot} = 2\pi^2 f_{rot}^2 I_{Pl}, \tag{6.4}$$

with f_{rot} denoting the rotation frequency and I_{PL} denoting the moment of inertia, which is ~ 16.6 kgm^2 for rotation around the main body z-axis (see Fig. 2.7), ~ 15.9 kgm^2 for rotation around the front landing gear x-axis and ~ 14.6 kgm^2 along the y-axis parallel to the lander balcony (Biele et al. 2015, Roll et al. 2016). During the collision the rotation pattern of the lander changed significantly, from a simple rotation along the z-axis (perpendicular to the landing gear plane) with a frequency of $f_1 = (77.0 \pm 1.0)$ mHz to a precession with a frequency of $f_2 = (42.0 \pm 1.0)$ mHz. After the second touchdown Philae had lost most of the rotational energy, spinning only with around $f_3 = (2.0 \pm 0.5)$ mHz. The flight times, the rotation rates, contact times, contact geometry and lander attitude used in the following are based on the results of section 4.3 and previous work by Heinisch et al. (2016). All these values are also summarized in Table 6.1. The resulting total velocity and energy distributions for the collision and second touchdown are depicted in Fig. 6.3

During the collision no significant acceleration along the z-axis perpendicular to the landing gear was detected, in contrast to the other touchdowns. This constraints the direction of the contact force to the landing gear plane. As the acceleration vector was approximately orthogonal to the lander z-axis and in line with the landing gear during contact,

	Collision	Second Touchdown
v_I	(0.33 ± 0.01) m/s	(0.22 ± 0.01) m/s
v_O	(0.22 ± 0.01) m/s	(0.05 ± 0.01) m/s
f_0	(77.0 ± 1.0) mHz	(42.0 ± 1.0) mHz
f_1	(42.0 ± 1.0) mHz	(2.0 ± 0.5) mHz
E_0	(7.02 ± 0.37) J	(2.97 ± 0.15) J
E_1	(2.83 ± 0.22) J	(0.10 ± 0.02) J
t	(2756 ± 3) s	(3926 ± 3) s
τ_c	(1.5 ± 1.0) s	(4.0 ± 1.0) s
A	(225 ± 40) cm^2	(350 ± 75) cm^2
σ	(399 ± 393) Pa	(147 ± 77) Pa

Table 6.1: Summary of the underlying values and results for the collision and second touchdown: total incoming and outgoing velocity v_I, v_O; Philae rotation rate before and after contact f_0, f_1; total energy before and after contact E_0, E_1; incoming flight time t, contact time τ_c, contact area A and resulting compressive strength σ.

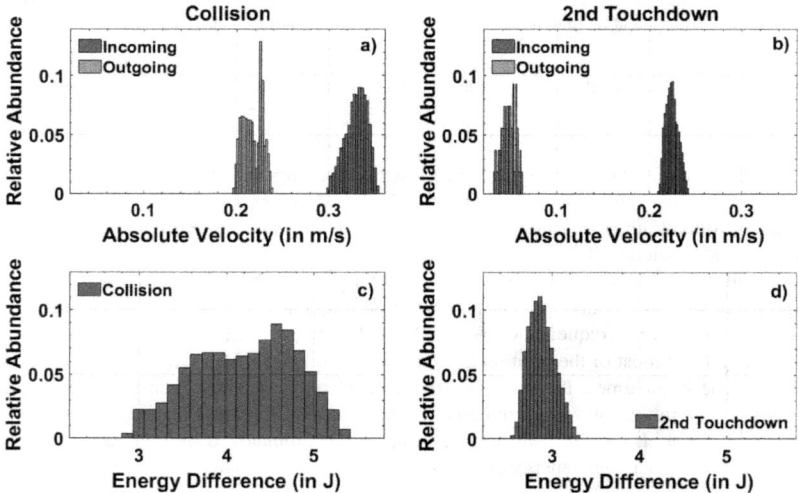

Figure 6.3: Histogram of the incoming and outgoing total velocities for the collision (panel a) and second touchdown (panel b) calculated based on each possible trajectory using a spatial resolution of 1 m. Panels c) and d) show the difference in energy before and after the collision and second touchdown respectively, which is equal to the work done during theses contacts (based on Heinisch et al. 2018a).

Philae must have struck the comet with the side structure of the foot assembly and not with the soles of the feet. This is the major difference between this collision and the other touchdowns and the reason that Auster et al. (2015) introduced the different naming scheme. The approximation of the direction of acceleration was made possible by the fact that the ROMAP magnetometer was mounted on the end of a spring supported boom. An assembly like this causes the magnetometer to swing from its nominal position as soon as the lander is accelerated perpendicular to the hinge axis. As the eigen frequency of the boom assembly is significantly above the excitation caused by the surface contacts, the boom movement correlates with the impact acceleration. As the background magnetic field is known (see section 3.2), this movement can be tracked using the magnetic field measurements, turning ROMAP into a simple accelerometer.

The attitude just after second touchdown implies that the balcony was pointing toward the comet and at least one foot, the tip of the ROMAP boom, and possibly part of the balcony was in contact with the surface. This was confirmed by a deflection of the ROMAP boom away from the landing gear and the fact that the ROMAP plasma sensor was covered by dust (see section 4.3.2). The position of the entrances of the plasma monitor means that the sensor head must have penetrated the surface at least ~ 5 cm for the sensor entrances to become covered by dust. This contact caused an upward deflection of the boom, as it is only supported by springs. As the plasma sensors housekeeping data remained nominal for the entire time, the sensor could not have sustained damage during contact. Especially the delicate charged retarding grids at the sensor entrance were not damaged. Considering an impact velocity of ≈ 0.22 m/s this is already a qualitative indication, that the surface material has to be soft.

6.2.1   Compressive Strength

The rebound of Philae offers the possibility of determining the compressive strength σ, which can be described as the threshold between plastic and elastic deformation. If the applied load is below the compressive strength limit of the material for a given area, the contact will be virtually elastic with negligible loss of kinetic energy leading to rebounding. This would cause a strong but transient acceleration. Otherwise, the contact becomes inelastic as a result of plastic deformation or reordering of the material (e.g. Schräpler et al. (2015)). As an inelastic contact is linked to a penetration of the material, the acceleration is weaker but spread out over a longer interval. Thus an upper estimate can be derived by examining the load on the cometary material applied during contact and change in kinetic energy linked to it. Granular material behaves differently under load compared to regular solids (Biele et al. 2009), for example exhibiting slip-stick interaction (Cole and Peters 2007). Therefore standard elastoplastic models (i.e. Yigit et al. 2011, Li et al. 2001) cannot be used. As there is no general description of the behavior of granular material under load (Cole and Peters 2007) a constant deceleration and hence a constant breaking force during contact is assumed. Even though this is not an exact description of real granular cometary material, it generally leads to an overestimation of the compressive strength, which is of no concern as the primary interest is in an upper bound. Furthermore additional energy losses inside the lander structure are neglected (similar on Biele et al. 2015). With these assumptions, an upper bound for the compressive strength of the material can be derived based on the total loss of energy using

$$\sigma = \frac{F}{A} = \frac{a_d \cdot m_{Pl}}{A} = \frac{\left(\sqrt{\frac{2E_0}{m_{Pl}}} - \sqrt{\frac{2E_1}{m_{Pl}}}\right) m_{Pl}}{A\tau_c}, \qquad (6.5)$$

Here F is the breaking force linked to the deceleration a_d. E_0 and E_1 denote the total energy before and after impact respectively, τ_c is the contact duration and A is the area of contact between the landing gear and the surface. The contact duration τ_c was determined from the movement of the ROMAP magnetometer boom caused by the acceleration during contact. As the boom is only supported by two springs, it can be deflected if an external acceleration is applied. This movement creates a characteristic signature in the observed magnetic field, which can be used to derive the duration of the acceleration, as explained in section 2.5.2. The area of contact A, was determined based on the known geometry of the landing gear, particularly the feet, combined with the angle of attack of the lander relative to the surface (derived from lander attitude) just prior to contact. The parameter $m_{Pl} = 97.63$ kg (Biele et al. 2015) is the mass of Philae. As the primary interest is in an upper bound for σ, it is assumed that the change in total kinetic ($E = E_{rot} + E_{kin}$) energy is only linked to the surface interaction. Hence, the deceleration is determined based on the total loss of energy during contact and not just the translational velocity.

For the collision event the total energy E_0 before contact was (7.02±0.37) J, which decreased to $E_1 = (2.83 \pm 0.22)$ J afterwards. Based on the reconstructed attitude (see section 4.3.1), Philae struck the surface with only one foot. In contrast to the 1st touchdown, the foot hit the comet not with the bottom of the soles but with the side of the foot structure, which leads to a contact area of $A = (225 \pm 40)$ cm^2. This area was estimated based on the CAD model of the foot assembly (depicted in Fig. 2.6) and a plane angled relative to the foot to represent the attitude of the lander. Even though the exact angle between the local surface and the foot is unknown, angular deviations only lead to small changes in the area of contact, because of the specifics of the foot design. The contact time $\tau_c = (1.5 \pm 1.0)$ s was determined from the acceleration derived from ROMAP observations, as explained above. This leads to an upper limit for the compressive strength of $\sigma = (399 \pm 393)$ Pa.

The same approach was used for the second touchdown. The kinetic energy E_0 before contact was (2.97±0.15) J, which is slightly higher than directly after the collision due to the difference in height between the area of the collision and the second touchdown. Afterward, the energy dropped to $E_1 = (0.10 \pm 0.02)$ J. The different contact geometry (at least one foot and magnetometer boom) leads to a higher contact area of (350 ± 75) cm^2 and a longer contact duration of $\tau_c = (4 \pm 1)$ s compared to the collision. This results in a compressive strength $\sigma = (147 \pm 77)$ Pa. These results are summarized in Table 6.1.

Images of Philae at the final Abydos landing site (Fig. 6.4) revealed several scratch features (S_A, S_B and S_C). Scratches S_A and S_B are located to the left of the lander and have a length of ~ 1.8 m, a width of ~ 11 cm and are ~ 42 cm apart. The length was derived from the number of pixels while accounting for projection effects and using the known dimensions of the lander for validation. It is possible that the lengths of S_A and S_B are slightly underestimated, as the boulder is only visible up to the edge of what is

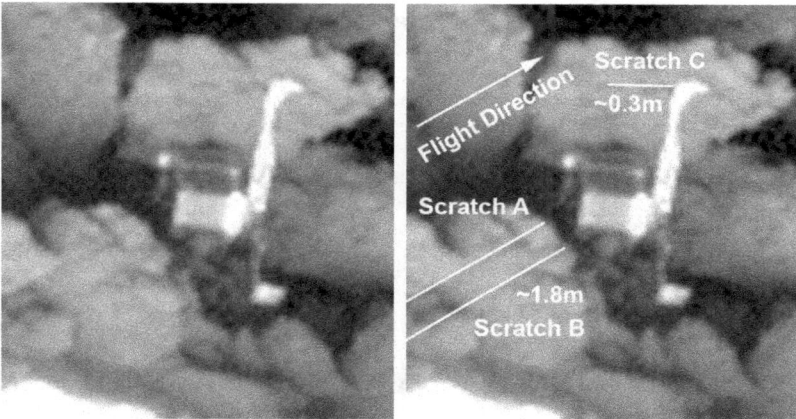

Figure 6.4: OSIRIS camera image of the Philae lander at the final Abydos landing site. The right panel is annotated to show the scratch marks 'A', 'B' and 'C', the approximated incoming flight direction is depicted by a white arrow. The image resolution is approximately 7 mm/px (based on Heinisch et al. 2018b).

depicted in Fig. 6.4. The imaging geometry means that some parts could be hidden behind smaller boulders in the foreground. Based on the size of these structures, a possible length error of 0.2 m was determined. The scratches are aligned with the incoming flight path of the lander coming from the second touchdown (see Fig. 6.2 for reference). As all high resolution images of this area were taken almost at the same time with similar geometry and lighting, it is impossible to reconstruct the exact 3D topography. Therefore the directionality of the scratch marks cannot be used to extrapolate the exact position of the second touchdown. Based on the available local terrain model it is only possible to estimate the approximate flight direction after the second touchdown. No other parallel structures at scales comparable to the scratch marks are visible in the images of the landing area (Bibring et al. 2015, Poulet et al. 2016), which suggests that the scratches were made by Philae. From the attitude information derived from combined magnetometer and housekeeping observations, as described in section 4.3.3, one can infer that two of the landing legs were pointing toward the surface shortly before the third touchdown (see illustration in Fig. 4.9). This makes it likely that the lander skidded across the surface for the final $\sim$ 1.8 m causing scratch marks S_A and S_B.

A third feature (scratch S_C) with a length of approximately 30 cm is visible above Philae next to one of the feet. It is assumed that scratch S_C was created while Philae came to a stop after it cleared the cliff (bottom left hand corner in Fig. 6.4) and tilted while hitting the wall it now rests on. As the direction of scratch S_C is parallel to the lander z-axis, the deceleration caused an observable movement of the ROMAP boom, which was visible in the magnetic field observations from 17:31:16 $\pm$ 1 s until 17:31:26 $\pm$ 1 s UTC.

The width of the scratches and the size of Philae's leg structure indicates, that only the soles of the feet have been in contact with the ground. This constrains the maximum penetration depth to approximately $\sim$ 5 cm (see Fig. 2.6 for dimensions of the foot), which is half the diameter of the soles. Otherwise the significantly larger upper foot structure (see Fig. 6.4 for total scale of the landing gear) would have been in contact with the surface causing substantial drag and considerably larger scratch marks in the range of $\sim$ 25 cm. A width-to-depth ratio of $\sim$ 2 is also consistent with other depressions found on 67P (El-Maarry et al. 2015) and can be considered realistic for the surface material. Iit is impossible to derive the actual depth of the scratches from the images because of the diffuse lighting conditions. From the characteristics of the scratches it is possible to estimate the contact pressure σ analogous to equation 6.5 based on the mechanical work done during creation of the scratches:

$$\sigma = \frac{F}{A} = \frac{\Delta E}{A l_s}. \tag{6.6}$$

Here A is again the area of contact, F is the breaking force, ΔE denotes the mechanical work done while creating the scratches and l_s is the length of the scratches. While the material in front of the foot is compressed, additional energy is required to overcome the shear and tensile strength (see i.e. Akono and Ulm (2011) for scratch modeling) at the boundary of the cometary material and the outer rim of the foot. As the aim of this study is to derive an upper limit for the compressive strength, it was assumed that the entire energy lost during the creation of the scratches was due to compression, hence leading to the highest possible compression force and preventing underestimation of the compressive strength.

To determine ΔE for scratches S_A and S_B, the remaining velocity afterwards was estimated as half of the mean initial velocity (again assuming constant deceleration) as a conservative first order approximation, based on the impact drag force analysis performed by Katsuragi and Blum (2017). The overall ratio of the individual values means that minor changes in the velocities would not cause major changes in the resulting compressive strength.

$$v_{out} = \frac{l_s}{2\tau_{AB}} \tag{6.7}$$

with $\tau_{AB} = (36 \pm 5)$ s and $l_s = (1.8 \pm 0.2)$ m, resulting in $v_{out} = (0.03 \pm 0.006)$ m/s. This yields a remaining kinetic energy of the lander of (0.04 ± 0.01) J, which combined with an incoming kinetic energy of (0.10 ± 0.02) J (see above) translates to $\Delta E = (0.06 \pm 0.03)$ J. With a contact area of $A_{AB} = (41 \pm 10)$ cm^2 estimated based on the foot geometry and penetration depth, this yields an upper limit for the compressive strength of $\sigma =(8 \pm 7)$ Pa. Because of the change in attitude and single foot contact while scratch S_C was created, the contact area was smaller with $A_C = (20 \pm 10)$ cm^2. With the remaining energy of (0.04 ± 0.01) J lost during contact and a length of the scratch of $l_s = (0.3 \pm 0.1)$ m, the resulting limit for the compressive strength is $\sigma = (73 \pm 70)$ Pa. These results are summarized in Table 6.2.

The force necessary to plow through the material depends on several material properties. Especially because of the low gravitational compression of the material, the resulting pressure is dominated by the compressive properties of the material by a factor of roughly ten especially (e.g. Groussin et al. 2015).

	Scratches A, B	Scratch C
E_0	(0.10 ± 0.02) J	(0.04 ± 0.01) J
E_1	(0.04 ± 0.01) J	0 J
l_s	(1.80 ± 0.20) m	(0.30 ± 1.0) m
A	(41 ± 10) cm^2	(20 ± 10) cm^2
σ	(8 ± 7) Pa	(73 ± 70) Pa

Table 6.2: Summary of the underlying values and results for scratches A,B and C: total energy before and after contact E_0, E_1; scratch length l_s, contact area A and resulting compressive strength σ.

6.3   Constraints on Comet Formation

The presented in situ results for the compressive strength span a range between 7 Pa and 399 Pa, which agrees well with the compressive strength of 30 Pa to 150 Pa estimated from remote observations by Groussin et al. (2015). Deviations between the upper limits for the compressive strength values derived for the individual contacts can most likely be attributed to local material variations. While the comet is homogeneous on larger sales, local differences in insolation, the dust-to-ice ratio (due to different outgassing rates, e.g. Lai et al. 2018) or volume filling factor are expected and lead to local deviations in the compressive strength (e.g. Fornasier et al. 2018). Possible uncertainties in the area of contact between the lander and the surface and errors in the estimated trajectories can also contribute to the differences, but these deviations should be covered by the conservative error range. Owing to the simplified mechanical model and the limited range of available measurements, the resulting compressive strength is only a local upper limit, without taking higher order mechanical characteristics of granular material into account.

Based on the mechanical interaction during the surface contacts, a compressive pressure below 100 Pa is already enough to cause significant compaction of the surface material. This compaction lead to an increase in compressive strength, which caused Philae to rebound at $\sim$ 160 Pa. This kind of material makes comets like 67P an ideal target for lander missions as the surface is strong enough to support the weight of a probe without overpenetration, but also soft enough to dissipate the kinetic energy during touchdown.

The range for the compressive strength derived above is slightly below the value of 1 kPa estimated by Biele et al. (2015) from the first touchdown, which can largely be attributed to the difference in dust coverage. In contrast to the first landing site Agilkia, neither the cliff near the Hatmehit crater nor Abydos have significant surface dust layers (see Fig. 6.1; Schröder et al. (2017), Keller et al. (2015), Thomas et al. (2015)). Hence, the mechanics of the subsequent contacts were not governed by surface dust layers and therefore provide insight into the properties of the more consolidated surface regions. This is also the reason why the scratch marks at Abydos are still visible two years after Philae landed and shows that regions that might look consolidated can in reality be extremely soft.

Based on the hammering of the MUPUS instrument, Spohn et al. (2015) postulated a compressive strength of 2 MPa for the Abydos site, which is significantly above the independently derived results of Biele et al. (2015) and Groussin et al. (2015) or the range presented above. A possible explanation for this discrepancy is an overestimation of the compressive strength caused by incorrect MUPUS deployment. The placement of the hammer could not be confirmed independently by camera images as initially intended. Hence, it is possible that the MUPUS penetrator toppled over after the deployment mechanism was retracted, for example due to a soft surface layer or uneven terrain. This would explain, why no further surface penetration was detected by the instrument after initial deployment on the surface. Even the highest hammer impact setting did not cause measurable penetration (Spohn et al. 2015). While this could be explained by an extremely hard surface as postulated by Spohn et al. (2015), it is also possible, that the MUPUS instrument toppled over and was parallel to the surface. A detailed analysis of the images of Philae at the final landing site revealed a bulge resembling the MUPUS hammer in shape and size below the front leg (+x leg) of Philae as shown in Fig. 6.5. No other components with similar dimension are mounted to the landing gear (see Fig. 2.5 as reference) and while the reflectivity is consistent with the casing of MUPUS, it does not match the surrounding terrain. This would also be consistent with the cable visible below the lander on ROLIS images, which was attributed to MUPUS by Schröder et al. (2017). Nominally this cable should lead away from the balcony to the MUPUS instrument deployed behind the balcony and therefore not be visible below the main lander body. In case MUPUS fell over and moved down across the boulder to below the landing gear leg, either due to gravity or the vibrations of the hammer mechanism, the cable would lead from the balcony directly down across the ROLIS field of view to the front of the lander, explaining the image.

The results of the scratch mark analysis are also inconsistent with a solidified surface ice layer with a thickness between 10 cm and 50 cm as proposed by Knapmeyer et al. (2018). It is also questionable, if such an ice layer is consistent with the fact, that the ROMAP plasma sensor was covered by dust during the second touchdown. This requires a relatively soft upper layer, as the ROMAP instrument was not damaged, but capable of breaking up the upper surface. As the results of Knapmeyer et al. (2018) rely on the correct deployment of the MUPUS hammer, uncertainties in the MUPUS operation might have caused an overprediction of the thickness of a possible surface ice layer. Several laboratory experiments have been performed to determine the mechanical properties of possible building blocks of comets. In particular Güttler et al. (2009), Schräpler et al. (2015), Lorek et al. (2016) and Katsuragi and Blum (2017) investigated how the compressibility of different dust structures created by random ballistic deposition as well as pebble sub-structures (with scales of centimeters, milimeters and 0.1 mm) and ice layers depend on the volume filling factor. The respective results are summarized in Fig. 6.6. Remote observations lead to multiple independent estimations for the volume filling factor. Kofman et al. (2015) derived values of 0.15...0.25, Pätzold et al. (2016) estimated 0.25...0.30 and a study by Fulle et al. (2016) resulted in 0.21...0.37. The corresponding range is depicted in Fig. 6.6 by the horizontal bar shaded in blue. As the volume filling factor is known, the laboratory measurements can be used to determine the most likely material of the surface. For the observed volume filling factor, dust and ice layers have a

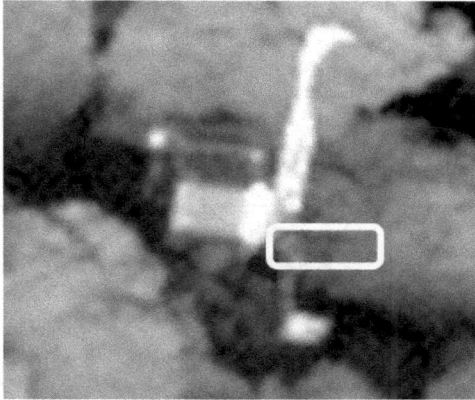

Figure 6.5: Closeup image of the final Philae landing site. A structure below the landing gear resembling the MUPUS penetrator was highlighted (adapted from Sierks 2016).

compressive strength above 10^3 Pa. A surface made up of such dust or ice layers is therefore inconsistent with the results of Groussin et al. (2015) or the compressive strength derived as part of this work.

In contrast, a surface consisting of layers of dust and ice aggregates could explain all observations, except the MPa range derived by Spohn et al. (2015). This is strong evidence for the presence of aggregate layers on the surface of 67P, i.e., a surface composed of sub-decimeter-sized aggregates as inferred from the gravitational collapse scenario. Such a conclusion can be drawn without precise knowledge of the actual value of the compressive strength, as the upper limit derived as part of this work and by Groussin et al. (2015), even considering all possible errors, is well below 10^3 Pa for different parts of 67P. The presence of such aggregate layers can only be explained by a formation dominated by relatively low velocity collisions. Hierarchical agglomeration would cause a significantly higher compressive strength due to impact-compaction of the pebbles and possible break-up of aggregates (Blum 2018). Hence, the derived mechanical properties combined with the available laboratory models suggest, that the gentle collapse of aggregate ensembles played a major role in the formation of comets like 67P.

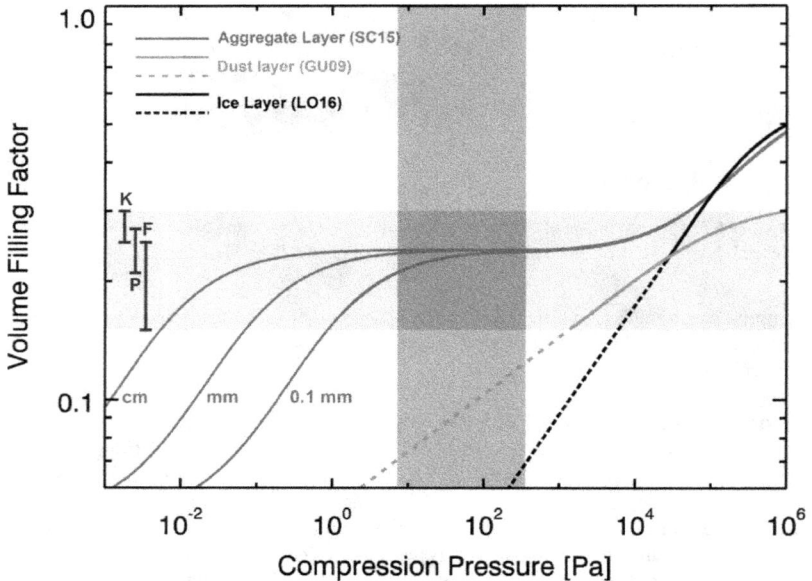

Figure 6.6: Dependence between volume filling factor and compressive pressure for aggregate, dust and water ice layers measured in the laboratory (SC15: Schräpler et al. 2015, GU09: Güttler et al. 2009, LO16: Lorek et al. 2016). The experimental results for dust and ice layers have been extrapolated for lower volume filling factors (dashed lines). The derived compression range for the surface contacts of Philae are depicted by the vertical red area. Measurements from Kofman et al. (2015) (K), Pätzold et al. (2016) (P) and Fulle et al. (2016) (F) have been used to determine the possible range for the volume filling factor (based on Heinisch et al. 2018b).

7 Summary and Outlook

This work describes how the magnetic field measurements from Philae's magnetometer ROMAP were used to further the understanding of the plasma environment and magnetization of comet 67P/Churyumov-Gerasimenko, as well as its surface properties and process of creation. To allow for the scientific interpretation of the measurements, two-point magnetic field observations were used to determine the attitude and dynamics of the Philae lander during descent and rebound and provide additional technical information about internal systems.

By combining images of Philae with ROMAP measurements, it was possible to use the lander as an impact probe and analyze the surface contacts to derive an overall upper limit of ~ 800 Pa for the compressive strength of the cometary surface. Coupled with previous laboratory results, the derived compressive strength is strong evidence for the presence of aggregate layers on the surface of 67P, that is, a surface composed of sub-decimeter-sized aggregates, as inferred from the gravitational collapse scenario (Davidsson et al. 2016, Blum 2018). The presence of such aggregate layers can only be explained by a formation dominated by relatively low-velocity collisions. The derived mechanical properties combined with the available laboratory models (Schräpler et al. 2015) therefore suggest, that the gentle collapse of aggregate ensembles played a major role in the formation of comets like 67P. As this information is crucial to determine what comets are made of and how they formed, it was one of the primary science goals of the Philae lander. Observational results for the mechanical properties are also vital for future laboratory studies, trying to explain the mechanisms behind outburts and the ejection of dust from the surface. As plans for different cometary sample return missions are being discussed, the compressive strength is also the most important parameter for the design of the sampling instruments.

The attitude reconstructed from the magnetic field observations also allowed for a detailed two-point analysis of low frequency magnetic singing comet waves (Richter et al. 2015). The wave frequency spans from 5 mHz to 50 mHz with an average wavelength of approximately 660 km. Based on a minimum-variance analysis, these waves propagate predominantly towards the Sun with a mean phase velocity of 5.3 km/s. A comparison between the dominant wave frequency, the overall power in the singing comet frequency range and the neutral gas pressure revealed, that in general a higher dominant frequency is linked to lower power and neutral pressure. Higher neutral densities cause an increase in wave power. The results of this work provide both qualitative and quantitative constraints for the singing comet waves. Hence they form the observational basis for future studies, especially numerical simulations, trying to understand the evolution of the plasma environment.

Based on the performed analysis, a new upper limit of 0.9 nT for the observed magnetic field on the surface of 67P was derived for any contribution from surface magnetization, significantly lowering the previous limit of 2 nT (Auster et al. 2015). The spatial resolution was also improved from ~1 m (Auster et al. 2015) to ~5 cm, which is in the range of typical aggregates. Assuming homogeneously magnetized pebbles in this size range, this translates to an upper limit of ~5 · 10^{-5} Am^2/kg for the specific magnetic moment. Based on models of the magnetization process (Fu and Weiss 2012, Biersteker et al. 2018) and depending on the exact history of formation, the absence of significant magnetization down to scales of ~5 cm, puts an upper threshold of 4 μT on the magnitude of the magnetic field in the solar nebular during the formation of comet 67P.

Apart from providing scientific in situ observations, the ROMAP measurements were also used to derive important information about the lander dynamics and attitude and analyze the operation of internal electrical systems, like the harpoon firing mechanism. Therefore this work showcased, that the versatility of magnetometers makes them a vital asset for space missions, especially if no dedicated housekeeping instruments are available. This was made possible in part by the availability of concurrent magnetic field observations from the orbiter magnetometer RPC-MAG. These measurements provided a magnetic field reference, allowing to discern between variations in the global magnetic environment and local changes linked to movement of the ROMAP sensors or interference from lander systems. As such this work highlights the importance of separate magnetometers on multi-spacecraft mission and showcases the advantages, especially for non nominal mission scenarios.

Bibliography

Acton, C. H., 1996, Ancillary data services of NASA's Navigation and Ancillary Information Facility , Planetary and Space Science, 44, 65 – 70, ISSN 0032-0633, planetary data system

Akono, A.-T., Ulm, F.-J., 2011, Scratch test model for the determination of fracture toughness, Engineering Fracture Mechanics, 78, 334 – 342, ISSN 0013-7944

Alfvèn, H., 1981, Cosmic Plasma (Astrophysics and Space Science Library), Springer, ISBN 9027711518

Alfvén, H., 1957, On the theory of comet tails, Tellus, 9, 92 – 96

Arnold, W., Faber, C., Knapmeyer, M., Witte, L., Schröder, S., Tune, J., Möhlmann, D., Roll, R., Chares, B., Fischer, H., Seidensticker, K., 2014, Inverting Comet Acoustic Surface Sounding Experiment (CASSE) touchdown signals to measure the elastic modulus of comet material, in: Asteroids, Comets, Meteors 2014, (Eds.) K. Muinonen, A. Penttilä, M. Granvik, A. Virkki, G. Fedorets, O. Wilkman, T. Kohout

Ashman, M., Barthelemy, M., O'Rourke, L., Almeida, M., Altobelli, N., Sitjà, M. C., Beteta, J. J. G., Geiger, B., Grieger, B., Heather, D., Hoofs, R., Küppers, M., Martin, P., Moissl, R., Crego, C. M., Pérez-Ayúcar, M., Suarez, E. S., Taylor, M., Vallat, C., 2016, Rosetta science operations in support of the Philae mission, Acta Astronautica, 125, 41 – 64, ISSN 0094-5765

Auster, H.-U., Apathy, I., Berghofer, G., Remizov, A., Roll, R., Fornacon, K., Glassmeier, K., Haerendel, G., Hejja, I., Kührt, E., Magnes, W., Moehlmann, D., Motschmann, U., Richter, I., Rosenbauer, H., Russell, C., Rustenbach, J., Sauer, K., Schwingenschuh, K., Szemerey, I., Waesch, R., 2007, ROMAP: Rosetta Magnetometer and Plasma Monitor, Space Science Reviews, 128, 221 – 240, ISSN 0038-6308

Auster, H.-U., Apathy, I., Berghofer, G., Fornacon, K.-H., Remizov, A., Carr, C., Güttler, C., Haerendel, G., Heinisch, P., Hercik, D., Hilchenbach, M., Kührt, E., Magnes, W., Motschmann, U., Richter, I., Russell, C. T., Przyklenk, A., Schwingenschuh, K., Sierks, H., Glassmeier, K.-H., 2015, The nonmagnetic nucleus of comet 67P/Churyumov-Gerasimenko, Science, 349, ISSN 0036-8075

Auster, H.-U. I. Richter, I., Glassmeier, K.-H., Berghofer, G., Carr, C., Motschmann, U., 2010, Magnetic field investigations during ROSETTA's 2867 Šteins flyby, Planetary and Space Science, 58, 1124 – 1128, ISSN 0032-0633, Special Issue: Rosetta Fly-by at Asteroid (2867) Steins

Balsiger, H., Altwegg, K., Bochsler, P., Eberhardt, P., Fischer, J., Graf, S., Jäckel, A., Kopp, E., Langer, U., Mildner, M., Müller, J., Riesen, T., Rubin, M., Scherer, S., Wurz, P., Wüthrich, S., Arijs, E., Delanoye, S., Keyser, J. D., Neefs, E., Nevejans, D., Rème, H., Aoustin, C., Mazelle, C., Médale, J.-L., Sauvaud, J. A., Berthelier, J.-J., Bertaux, J.-L., Duvet, L., Illiano, J.-M., Fuselier, S. A., Ghielmetti, A. G., Magoncelli, T., Shelley, E. G., Korth, A., Heerlein, K., Lauche, H., Livi, S., Loose, A., Mall, U., Wilken, B., Gliem, F., Fiethe, B., Gombosi, T. I., Block, B., Carignan, G. R., Fisk, L. A., Waite, J. H., Young, D. T., Wollnik, H., 2007, Rosina – rosetta orbiter spectrometer for ion and neutral analysis, Space Science Reviews, 128, 745 – 801, ISSN 1572-9672

Baranyai, T., Balázs, A., Várkonyi, P. L., 2016, Partial reconstruction of the rotational motion of Philae spacecraft during its landing on comet 67P/Churyumov-Gerasimenko, ArXiv e-prints, 1604.04414

Barthelemy, M., Heather, D., Grotheer, E., Besse, S., Andres, R., Vallejo, F., Barnes, T., Kolokolova, L., O'Rourke, L., Fraga, D., A'Hearn, M., Martin, P., Taylor, M., 2018, Rosetta: How to archive more than 10 years of mission, Planetary and Space Science, 150, 91 – 103, ISSN 0032-0633

Barucci, M. A., Fulchignoni, M., Rossi, A., 2007, Rosetta Asteroid Targets: 2867 Steins and 21 Lutetia, Space Science Reviews, 128, 67 – 78, ISSN 1572-9672

Behar, E., Nilsson, H., Alho, M., Goetz, C., Tsurutani, B., 2017, The birth and growth of a solar wind cavity around a comet – Rosetta observations, Monthly Notices of the Royal Astronomical Society, 469, 396 – 403

Benesty, J., Chen, J., Huang, Y., Cohen, I., 2009, Pearson Correlation Coefficient, pp. 1 – 4, Springer Berlin Heidelberg, Berlin, Heidelberg, ISBN 978-3-642-00296-0

Berger, E. L., Zega, T. J., Keller, L. P., Lauretta, D. S., 2011, Evidence for aqueous activity on comet 81P/Wild 2 from sulfide mineral assemblages in stardust samples and ci chondrites, Geochimica et Cosmochimica Acta, 75, 3501 – 3513, ISSN 0016-7037

Bibring, J.-P., Lamy, P., Langevin, Y., Soufflot, A., Berthé, M., Borg, J., Poulet, F., Mottola, S., 2007a, CIVA, Space Science Reviews, 128, 397 – 412, ISSN 0038-6308

Bibring, J.-P., Rosenbauer, H., Boehnhardt, H., Ulamec, S., Biele, J., Espinasse, S., Feuerbacher, B., Gaudon, P., Hemmerich, P., Kletzkine, P., Moura, D., Mugnuolo, R., Nietner, G., Pätz, B., Roll, R., Scheuerle, H., Szegö, K., Wittmann, K., 2007b, The Rosetta Lander Philae Investigations, Space Science Reviews, 128, 205 – 220, ISSN 0038-6308

Bibring, J.-P., Langevin, Y., Carter, J., Eng, P., Gondet, B., Jorda, L., Le Mouélic, S., Mottola, S., Pilorget, C., Poulet, F., Vincendon, M., 2015, 67P/Churyumov-Gerasimenko surface properties as derived from civa panoramic images, Science, 349, ISSN 0036-8075

Biele, J., Ulamec, S., Richter, L., Knollenberg, J., Kührt, E., Möhlmann, D., 2009, The putative mechanical strength of comet surface material applied to landing on a comet, Acta Astronautica, 65, 1168 – 1178, ISSN 0094-5765

Biele, J., Ulamec, S., Maibaum, M., Roll, R., Witte, L., Jurado, E., Muñoz, P., Arnold, W., Auster, H.-U., Casas, C., Faber, C., Fantinati, C., Finke, F., Fischer, H.-H., Geurts, K., Güttler, C., Heinisch, P., Herique, A., Hviid, S., Kargl, G., Knapmeyer, M., Knollenberg, J., Kofman, W., Kömle, N., Kührt, E., Lommatsch, V., Mottola, S., Pardo de Santayana, R., Remetean, E., Scholten, F., Seidensticker, K., Sierks, H., Spohn, T., 2015, The landing(s) of Philae and inferences about comet surface mechanical properties, Science, 349

Biermann, L., 1951, Kometenschweife und solare Korpuskularstrahlung, Zeitschrift für Astrophysik, 29, 274 – 286

Biermann, L., Brosowski, B., Schmidt, H. U., 1967, The interaction of the solar wind with a comet, Solar Physics, 1, 254 – 284, ISSN 1573-093X

Biersteker, J. B., Weiss, B. P., Heinisch, P., Herčik, D., Glassmeier, K.-H., Auster, H.-U., 2018, Constraints on Magnetic Field Intensity in the Outer Solar Nebula During Formation of Comet 67P/Churyumov-Gerasimenko from Philae Magnetometry, Lunar and Planetary Science Conference, 49, 2642

Blum, J., 2018, Dust evolution in protoplanetary discs and the formation of planetesimals, Space Science Reviews, 214, 52 1 – 52 19, ISSN 1572-9672

Blum, J., Gundlach, B., Mühle, S., Trigo-Rodriguez, J., 2014, Comets formed in solar-nebula instabilities! – an experimental and modeling attempt to relate the activity of comets to their formation process, Icarus, 235, 156 – 169, ISSN 0019-1035

Blum, J., Gundlach, B., Krause, M., Fulle, M., Johansen, A., Agarwal, J., von Borstel, I., Shi, X., Hu, X., Bentley, M. S., Capaccioni, F., Colangeli, L., Della Corte, V., Fougere, N., Green, S. F., Ivanovski, S., Mannel, T., Merouane, S., Migliorini, A., Rotundi, A., Schmied, R., Snodgrass, C., 2017, Evidence for the formation of comet 67P/Churyumov-Gerasimenko through gravitational collapse of a bound clump of pebbles, Monthly Notices of the Royal Astronomical Society, 469, 755 – 773

Buemi, M., Landi, A., Procopio, D., 2000, Autonomous Star Tracker for ROSETTA, in: Spacecraft Guidance, Navigation and Control Systems, (Ed.) B. Schürmann, vol. 425 of ESA Special Publication, p. 279

Burch, J. L., Goldstein, R., Cravens, T. E., Gibson, W. C., Lundin, R. N., Pollock, C. J., Winningham, J. D., Young, D. T., 2007, Rpc-ies: The ion and electron sensor of the rosetta plasma consortium, Space Science Reviews, 128, 697 – 712, ISSN 1572-9672

Butterworth, S., 1930, On the theory of filter amplifiers, Wireless Engineer, 7, 536 – 541

Carr, C., Cupido, E., Lee, C. G. Y., Balogh, A., Beek, T., Burch, J. L., Dunford, C. N., Eriksson, A. I., Gill, R., Glassmeier, K. H., Goldstein, R., Lagoutte, D., Lundin, R., Lundin, K., Lybekk, B., Michau, J. L., Musmann, G., Nilsson, H., Pollock, C., Richter,

I., Trotignon, J. G., 2007, RPC: The Rosetta Plasma Consortium, Space Science Reviews, 128, 629 – 647, ISSN 1572-9672

Colangeli, L., Lopez-Moreno, J. J., Palumbo, P., Rodriguez, J., Cosi, M., Corte, V. D., Esposito, F., Fulle, M., Herranz, M., Jeronimo, J. M., Lopez-Jimenez, A., Epifani, E. M., Morales, R., Moreno, F., Palomba, E., Rotundi, A., 2007, The Grain Impact Analyser and Dust Accumulator (GIADA) Experiment for the Rosetta Mission: Design, Performances and First Results, Space Science Reviews, 128, 803 – 821, ISSN 1572-9672

Cole, D. M., Peters, J. F., 2007, A physically based approach to granular media mechanics: grain-scale experiments, initial results and implications to numerical modeling, Granular Matter, 9, 309 – 321, ISSN 1434-7636

Coradini, A., Capaccioni, F., Drossart, P., Arnold, G., Ammannito, E., Angrilli, F., Barucci, A., Bellucci, G., Benkhoff, J., Bianchini, G., Bibring, J. P., Blecka, M., Bockelee-Morvan, D., Capria, M. T., Carlson, R., Carsenty, U., Cerroni, P., Colangeli, L., Combes, M., Combi, M., Crovisier, J., Desanctis, M. C., Encrenaz, E. T., Erard, S., Federico, C., Filacchione, G., Fink, U., Fonti, S., Formisano, V., Ip, W. H., Jaumann, R., Kuehrt, E., Langevin, Y., Magni, G., Mccord, T., Mennella, V., Mottola, S., Neukum, G., Palumbo, P., Piccioni, G., Rauer, H., Saggin, B., Schmitt, B., Tiphene, D., Tozzi, G., 2007, Virtis: An Imaging Spectrometer for the Rosetta Mission, Space Science Reviews, 128, 529 – 559, ISSN 1572-9672

D'Accolti, G., Beltrame, G., Ferrando, E., Brambilla, L., Contini, R., Vallini, L., Mugnuolo, R., Signorini, C., Fiebrich, H., Caon, A., 2002, The Solar Array Photovoltaic Assembly for the ROSETTA Orbiter and Lander Spacecraft's, in: Space Power, (Ed.) A. Wilson, vol. 502 of ESA Special Publication, pp. 445 – 451

Davidsson, B., Sierks, H., Güttler, C., Marzari, F., Pajola, M., Rickman, H., A´Hearn, M. F., Auger, A.-T., El-Maarry, M. R., Fornasier, S., Gutiérrez, P. J., Keller, H. U., Massironi, M., Snodgrass, C., Vincent, J.-B., Barbieri, C., Lamy, P. L., Rodrigo, R., Koschny, D., Barucci, M. A., Bertaux, J.-L., Bertini, I., Cremonese, G., Da Deppo, V., Debei, S., De Cecco, M., Feller, C., Fulle, M., Groussin, O., Hviid, S. F., Höfner, S., Ip, W.-H., Jorda, L., Knollenberg, J., Kovacs, G., Kramm, J.-R., Kührt, E., Küppers, M., La Forgia, F., Lara, L. M., Lazzarin, M., Lopez Moreno, J. J., Moissl-Fraund, R., Mottola, S., Naletto, G., Oklay, N., Thomas, N., Tubiana, C., 2016, The primordial nucleus of comet 67P/Churyumov-Gerasimenko, Astronomy & Astrophysics, 592, A63 1 – A63 30

Dudal, C., Loisel, C., 2016, Rosetta-Philae RF link, challenging communications from a comet, Acta Astronautica, 125, 137 – 148, ISSN 0094-5765

El-Maarry, M. R., Thomas, N., Giacomini, L., Massironi, M., Pajola, M., Marschall, R., Gracia-Berná, A., Sierks, H., Barbieri, C., Lamy, P. L., Rodrigo, R., Rickman, H., Koschny, D., Keller, H. U., Agarwal, J., A´Hearn, M. F., Auger, A.-T., Barucci, M. A., Bertaux, J.-L., Bertini, I., Besse, S., Bodewits, D., Cremonese, G., Da Deppo, V., Davidsson, B., De Cecco, M., Debei, S., Güttler, C., Fornasier, S., Fulle, M., Groussin, O., Gutierrez, P. J., Hviid, S. F., Ip, W.-H., Jorda, L., Knollenberg, J., Kovacs, G., Kramm, J.-R., Kührt, E., Küppers, M., La Forgia, F., Lara, L. M., Lazzarin,

M., Lopez Moreno, J. J., Marchi, S., Marzari, F., Michalik, H., Naletto, G., Oklay, N., Pommerol, A., Preusker, F., Scholten, F., Tubiana, C., Vincent, J.-B., 2015, Regional surface morphology of comet 67P/Churyumov-Gerasimenko from Rosetta/OSIRIS images, Astronomy & Astrophysics, 583, A26 1 – A26 28

Eriksson, A. I., Boström, R., Gill, R., Åhlén, L., Jansson, S.-E., Wahlund, J.-E., André, M., Mälkki, A., Holtet, J. A., Lybekk, B., Pedersen, A., Blomberg, L. G., The LAP Team, 2007, Rpc-lap: The rosetta langmuir probe instrument, Space Science Reviews, 128, 729 – 744, ISSN 1572-9672

Faber, C., Knapmeyer, M., Roll, R., Chares, B., Schröder, S., Witte, L., Seidensticker, K. J., Fischer, H.-H., Möhlmann, D., Arnold, W., 2015, A method for inverting the touchdown shock of the philae lander on comet 67P/Churyumov-Gerasimenko, Planetary and Space Science, 106, 46 – 55, ISSN 0032-0633

Ferri, P., Accomazzo, A., Hubault, A., Lodiot, S., Pellon-Bailon, J.-L., Porta, R., 2012, Rosetta enters hibernation, Acta Astronautica, 79, 124 – 130, ISSN 0094-5765

Festou, M. C., Rickman, H., West, R. M., 1993, Comets, The Astronomy and Astrophysics Review, 4, 363 – 447, ISSN 1432-0754

Finzi, A. E., Zazzera, F. B., Dainese, C., Malnati, F., Magnani, P. G., Re, E., Bologna, P., Espinasse, S., Olivieri, A., 2007, Sd2 – how to sample a comet, Space Science Reviews, 128, 281 – 299, ISSN 1572-9672

Fornasier, S., Hoang, V. H., Hasselmann, P. H., Feller, C., Barucci, M. A., Deshapriya, J. D. P., Sierks, H., Naletto, G., Lamy, P. L., Rodrigo, R., Koschny, D., Davidsson, B., Agarwal, J., Barbieri, C., Bertaux, J. L., Bertini, I., Bodewits, D., Cremonese, G., Deppo, V. D., Debei, S., Cecco, M. D., Deller, J., Ferrari, S., Fulle, M., Gutierrez, P. J., Guttler, C., Ip, W. H., Keller, H. U., Küppers, M., Forgia, F. L., Lara, M. L., Lazzarin, M., Lin, Z.-Y., Moreno, J. J. L., Marzari, F., Mottola, S., Pajola, M., Shi, X., Toth, I., Tubiana, C., 2018, Linking surface morphology, composition, and activity on the nucleus of 67P/Churyumov-Gerasimenko, Astronomy & Astrophysics, in press

Fu, R., Weiss, B., 2012, Detrital remanent magnetization in the solar nebula, Journal of Geophysical Research: Planets, 117, 1 – 19

Fu, R. R., Weiss, B. P., Schrader, D. L., Johnson, B. C., 2018, Records of Magnetic Fields in the Chondrule Formation Environment, pp. 324 — 340, Cambridge University Press

Fulle, M., Della Corte, V., Rotundi, A., Rietmeijer, F. J. M., Green, S. F., Weissman, P., Accolla, M., Colangeli, L., Ferrari, M., Ivanovski, S., Lopez-Moreno, J. J., Epifani, E. M., Morales, R., Ortiz, J. L., Palomba, E., Palumbo, P., Rodriguez, J., Sordini, R., Zakharov, V., 2016, Comet 67P/Churyumov-Gerasimenko preserved the pebbles that formed planetesimals, Monthly Notices of the Royal Astronomical Society, 462, 132 – 137

Garmier, R., Ceolin, T., Martin, T., Blazquez, A., Canalias, E., Jurado, E., Remetean, E., Lauren-Varin, J., Dolvies, B., Herique, A., Roger, Y., Kofman, W., Puget, P., Pasquero, P., Zine, S., Jorda, L., Heinisch, P., 2015, Philae Landing on Comet Churyumov-Gerasimenko: Understanding of Its Descent Trajectory, Attitude, Rebound and Final Landing Site, International Symposium on Space Flight Dynamics, 25

Gellert, R., Rieder, R., Brückner, J., Clark, B. C., Dreibus, G., Klingelhöfer, G., Lugmair, G., Ming, D. W., Wänke, H., Yen, A., Zipfel, J., Squyres, S. W., 2006, Alpha Particle X-Ray Spectrometer (APXS): Results from Gusev crater and calibration report, Journal of Geophysical Research: Planets, 111, E02S05 1 – E02S05 32

Glassmeier, K.-H., 2017, Interaction of the solar wind with comets: a rosetta perspective, Philosophical Transactions of the Royal Society of London A: Mathematical, Physical and Engineering Sciences, 375, ISSN 1364-503X

Glassmeier, K.-H., Neubauer, F. M., 1993, Low-frequency electromagnetic plasma waves at comet P/Grigg-Skjellerup: Overview and spectral characteristics, Journal of Geophysical Research: Space Physics, 98, 20 921 – 20 935

Glassmeier, K.-H., Coates, A. J., Acuña, M. H., Goldstein, M. L., Johnstone, A. D., Neubauer, F. M., Rème, H., 1989, Spectral characteristics of low-frequency plasma turbulence upstream of comet p/halley, Journal of Geophysical Research: Space Physics, 94, 37 – 48

Glassmeier, K.-H., Tsurutani, B., Neubauer, F., 1997, Adventures in parameter space a comparison of low-frequency plasma waves at comets In: Hada, T., Matsumoto, H. (Eds.), Nonlinear Waves and Chaos in Space Plasmas, Terra Scientific Publishing Company, Tokyo

Glassmeier, K.-H., Boehnhardt, H., Koschny, D., Kührt, E., Richter, I., 2007a, The Rosetta Mission: Flying Towards the Origin of the Solar System, Space Science Reviews, 128, 1 – 21, ISSN 0038-6308

Glassmeier, K.-H., Richter, I., Diedrich, A., Musmann, G., Auster, U., Motschmann, U., Balogh, A., Carr, C., Cupido, E., Coates, A., Rother, M., Schwingenschuh, K., Szegö, K., Tsurutani, B., 2007b, RPC-MAG The Fluxgate Magnetometer in the ROSETTA Plasma Consortium, Space Science Reviews, 128, 649 – 670, ISSN 0038-6308

Goesmann, F., Rosenbauer, H., Bredehöft, J. H., Cabane, M., Ehrenfreund, P., Gautier, T., Giri, C., Krüger, H., Le Roy, L., MacDermott, A. J., McKenna-Lawlor, S., Meierhenrich, U. J., Caro, G. M. M., Raulin, F., Roll, R., Steele, A., Steininger, H., Sternberg, R., Szopa, C., Thiemann, W., Ulamec, S., 2015, Organic compounds on comet 67P/Churyumov-Gerasimenko revealed by cosac mass spectrometry, Science, 349, ISSN 0036-8075

Goetz, C., Koenders, C., Richter, I., Altwegg, K., Burch, J., Carr, C., Cupido, E., Eriksson, A., Güttler, C., Henri, P., Mokashi, P., Nemeth, Z., Nilsson, H., Rubin, M., Sierks, H., Tsurutani, B., Vallat, C., Volwerk, M., Glassmeier, K.-H., 2016, First detection of a diamagnetic cavity at comet 67P/Churyumov-Gerasimenko, Astronomy & Astrophysics, 588, A24 1 – A24 6

Goldstein, M. L., Wong, H. K., Glassmeier, K. H., 1990a, Generation of low-frequency waves at comet halley, Journal of Geophysical Research: Space Physics, 95, 947 – 955

Goldstein, M. L., Wong, H. K., Glassmeier, K. H., 1990b, Generation of low-frequency waves at comet halley, Journal of Geophysical Research: Space Physics, 95, 947 – 955, ISSN 2156-2202

Grewing, M., Praderie, F., Reinhard, R. (Eds.), 1988, Exploration of Halley's Comet, Springer, Berlin Heidelberg

Groussin, O., Jorda, L., Auger, A.-T., Kührt, E., Gaskell, R., Capanna, C., Scholten, F., Preusker, F., Lamy, P., Hviid, S., Knollenberg, J., Keller, U., Huettig, C., Sierks, H., Barbieri, C., Rodrigo, R., Koschny, D., Rickman, H., A´Hearn, M. F., Agarwal, J., Barucci, M. A., Bertaux, J.-L., Bertini, I., Boudreault, S., Cremonese, G., Da Deppo, V., Davidsson, B., Debei, S., De Cecco, M., El-Maarry, M. R., Fornasier, S., Fulle, M., Gutiérrez, P. J., Güttler, C., Ip, W.-H, Kramm, J.-R., Küppers, M., Lazzarin, M., Lara, L. M., Lopez Moreno, J. J., Marchi, S., Marzari, F., Massironi, M., Michalik, H., Naletto, G., Oklay, N., Pommerol, A., Pajola, M., Thomas, N., Toth, I., Tubiana, C., Vincent, J.-B., 2015, Gravitational slopes, geomorphology, and material strengths of the nucleus of comet 67P/Churyumov-Gerasimenko from OSIRIS observations, Astronomy & Astrophysics, 583, A32 1 – A32 12

Grygorczuk, J., Dobrowolski, M., Wisniewski, L., Banaszkiewicz, M., Ciesielska, M., Kedziora, B., Seweryn, K., Wawrzaszek, R., Wierzchon, T., Trzaska, M., Ossowski, M., 2011, Advanced mechanisms and tribological tests of the hammering sampling device chomik, in: 14th European Space Mechanisms & Tribology Symposium – European Space Mechanisms and Tribology Symposium 2011

Gulkis, S., Frerking, M., Crovisier, J., Beaudin, G., Hartogh, P., Encrenaz, P., Koch, T., Kahn, C., Salinas, Y., Nowicki, R., Irigoyen, R., Janssen, M., Stek, P., Hofstadter, M., Allen, M., Backus, C., Kamp, L., Jarchow, C., Steinmetz, E., Deschamps, A., Krieg, J., Gheudin, M., Bockelée-Morvan, D., Biver, N., Encrenaz, T., Despois, D., Ip, W., Lellouch, E., Mann, I., Muhleman, D., Rauer, H., Schloerb, P., Spilker, T., 2007, Miro: Microwave instrument for rosetta orbiter, Space Science Reviews, 128, 561 – 597, ISSN 1572-9672

Güttler, C., Krause, M., Geretshauser, R. J., Speith, R., Blum, J., 2009, The physics of protoplanetesimal dust agglomerates. iv. toward a dynamical collision model, The Astrophysical Journal, 701, 130 1 – 130 14

Hahnel, R., Hegler, S., Statz, C., Plettemeier, D., Zine, S., Herique, A., Kofman, W., 2015, Consert line-of-sight link budget simulator, Planetary and Space Science, 111, 55 – 61, ISSN 0032-0633

Hajra, R., Henri, P., Vallières, X., Moré, J., Gilet, N., Wattieaux, G., Goetz, C., Richter, I., Tsurutani, B. T., Gunell, H., Nilsson, H., Eriksson, A. I., Nemeth, Z., Burch, J. L., Rubin, M., 2018, Dynamic unmagnetized plasma in the diamagnetic cavity around comet 67P/Churyumov-Gerasimenko, Monthly Notices of the Royal Astronomical Society, 475, 4140 – 4147

Halley, E., 1705, A synopsis of the astronomy of comets, Royal Society of London - Philosophical transactions

Hansen, K. C., Altwegg, K., Berthelier, J.-J., Bieler, A., Biver, N., Bockelée-Morvan, D., Calmonte, U., Capaccioni, F., Combi, M. R., De Keyser, J., Fiethe, B., Fougere, N., Fuselier, S. A., Gasc, S., Gombosi, T. I., Huang, Z., Le Roy, L., Lee, S., Nilsson, H., Rubin, M., Shou, Y., Snodgrass, C., Tenishev, V., Toth, G., Tzou, C.-Y., Simon Wedlund, C., the ROSINA team, 2016, Evolution of water production of 67P/Churyumov-Gerasimenko: an empirical model and a multi-instrument study, Monthly Notices of the Royal Astronomical Society, 462, 491 – 506

Hausman, B. A., Michel, F. C., Espley, J. R., Cloutier, P. A., 2004, On determining the nature and orientation of magnetic directional discontinuities: Problems with the minimum variance method, Journal of Geophysical Research: Space Physics, 109, A11 102 1 – A11 102 8

Heinisch, P., Auster, H.-U., 2015, Determination of variometer alignment by using variation comparison with DI3-flux, The Journal of Indian Geophysical Union, 19, 433 – 446

Heinisch, P., Auster, H.-U., Richter, I., Hercik, D., Jurado, E., Garmier, R., Güttler, C., Glassmeier, K.-H., 2016, Attitude reconstruction of ROSETTA's Lander PHILAE using two-point magnetic field observations by ROMAP and RPC-MAG, Acta Astronautica, 125, 174 – 182, ISSN 0094-5765

Heinisch, P., Auster, H.-U., Plettemeier, D., Kofman, W., Herique, A., Statz, C., Hahnel, R., Rogez, Y., Richter, I., Hilchenbach, M., Jurado, E., Garmier, R., Martin, T., Finke, F., Güttler, C., Sierks, H., Glassmeier, K.-H., 2017a, Reconstruction of the flight and attitude of rosetta's lander philae, Acta Astronautica, 140, 509 – 516, ISSN 0094-5765

Heinisch, P., Auster, H.-U., Richter, I., Haerendel, G., Apathy, I., Fornacon, K.-H., Cupido, E., Glassmeier, K.-H., 2017b, Joint two-point observations of lf-waves at 67P/Churyumov-Gerasimenko, Monthly Notices of the Royal Astronomical Society, 469, 68 — 72

Heinisch, P., Auster, H.-U., Gundlach, B., Blum, J., Güttler, C., Tubiana, C., Sierks, H., Hilchenbach, M., Biele, J., Richter, I., Glassmeier, K. H., 2018a, Compressive strength of comet 67P/Churyumov-Gerasimenko derived from philae surface contacts, Astronomy & Astrophysics, in press

Heinisch, P., Auster, H.-U., Richter, I., Glassmeier, K.-H., 2018b, Revisiting the magnetization of comet 67P/Churyumov-Gerasimenko, Astronomy & Astrophysics, in press

Heritier, K. L., Galand, M., Henri, P., Johansson, F. L., Beth, A., Eriksson, A. I. and Vallières, X., Altwegg, K., Burch, J. L., Carr, C., Ducrot, E., Hajra, R., Rubin, M., 2018, Plasma source and loss at comet 67p during the rosetta mission, Astronomy & Astrophysics, 618, A77 1 – A77 18

Israelevich, P. L., Ershkovich, A. I., 1994, Induced magnetosphere of comet halley: 2. magnetic field and electric currents, Journal of Geophysical Research: Space Physics, 99, 21 225 – 21 232

Johansen, A., Oishi, J. S., Low, M.-M. M., Klahr, H., Henning, T., Youdin, A., 2007, Rapid planetesimal formation in turbulent circumstellar disks, Nature, 448, 1022 — 1025

Jorda, L., Gaskell, R., Capanna, C., Hviid, S., Lamy, P., Ďurech, J., Faury, G., Groussin, O., Gutiérrez, P., Jackman, C., Keihm, S., Keller, H., Knollenberg, J., Kührt, E., Marchi, S., Mottola, S., Palmer, E., Schloerb, F., Sierks, H., Vincent, J.-B., A'Hearn, M., Barbieri, C., Rodrigo, R., Koschny, D., Rickman, H., Barucci, M., Bertaux, J., Bertini, I., Cremonese, G., Deppo, V. D., Davidsson, B., Debei, S., Cecco, M. D., Fornasier, S., Fulle, M., Güttler, C., Ip, W.-H., Kramm, J., Küppers, M., Lara, L., Lazzarin, M., Moreno, J. L., Marzari, F., Naletto, G., Oklay, N., Thomas, N., Tubiana, C., Wenzel, K.-P., 2016, The global shape, density and rotation of comet 67P/Churyumov-Gerasimenko from preperihelion rosetta/osiris observations, Icarus, 277, 257 – 278, ISSN 0019-1035

Jurado, E., Martin, T., Canalias, E., Blazquez, A., Garmier, R., Ceolin, T., Gaudon, P., Delmas, C., Biele, J., Ulamec, S., Remetean, E., Torres, A., Laurent-Varin, J., Dolives, B., Herique, A., Rogez, Y., Kofman, W., Jorda, L., Zakharov, V., Crifo, J.-F., Rodionov, A., Heinisch, P., Vincent, J.-B., 2016, Rosetta lander philae: Flight dynamics analyses for landing site selection and post-landing operations, Acta Astronautica, 125, 65 – 79, ISSN 0094-5765

Katsuragi, H., Blum, J., 2017, The physics of protoplanetesimal dust agglomerates. ix. mechanical properties of dust aggregates probed by a solid-projectile impact, The Astrophysical Journal, 851, 23 1 – 23 10

Keller, H. U., Barbieri, C., Lamy, P., Rickman, H., Rodrigo, R., Wenzel, K.-P., Sierks, H., A'Hearn, M. F., Angrilli, F., Angulo, M., Bailey, M. E., Barthol, P., Barucci, M. A., Bertaux, J.-L., Bianchini, G., Boit, J.-L., Brown, V., Burns, J. A., Büttner, I., Castro, J. M., Cremonese, G., Curdt, W., Deppo, V. D., Debei, S., Cecco, M. D., Dohlen, K., Fornasier, S., Fulle, M., Germerott, D., Gliem, F., Guizzo, G. P., Hviid, S. F., Ip, W.-H., Jorda, L., Koschny, D., Kramm, J. R., Kührt, E., Küppers, M., Lara, L. M., Llebaria, A., López, A., López-Jimenez, A., López-Moreno, J., Meller, R., Michalik, H., Michelena, M. D., Müller, R., Naletto, G., Origné, A., Parzianello, G., Pertile, M., Quintana, R., Ragazzoni, R., Ramous, P., Reiche, K.-U., Reina, M., Rodríguez, J., Rousset, G., Sabau, L., Sanz, A., Sivan, J.-P., Stöckner, K., Tabero, J., Telljohann, U., Thomas, N., Timon, V., Tomasch, G., Wittrock, T., Zaccariotto, M., 2007, Osiris – the scientific camera system onboard rosetta, Space Science Reviews, 128, 433 – 506, ISSN 1572-9672

Keller, H. U., Barbieri, C., Koschny, D., Lamy, P., Rickman, H., Rodrigo, R., Sierks, H., A'Hearn, M. F., Angrilli, F., Barucci, M. A., Bertaux, J.-L., Cremonese, G., Da Deppo, V., Davidsson, B., De Cecco, M., Debei, S., Fornasier, S., Fulle, M., Groussin, O., Gutierrez, P. J., Hviid, S. F., Ip, W.-H., Jorda, L., Knollenberg, J., Kramm, J. R., Kührt,

E., Küppers, M., Lara, L.-M., Lazzarin, M., Moreno, J. L., Marzari, F., Michalik, H., Naletto, G., Sabau, L., Thomas, N., Wenzel, K.-P., Bertini, I., Besse, S., Ferri, F., Kaasalainen, M., Lowry, S., Marchi, S., Mottola, S., Sabolo, W., Schröder, S. E., Spjuth, S., Vernazza, P., 2010, E-Type Asteroid (2867) Steins as Imaged by OSIRIS on Board Rosetta, Science, 327, 190 – 193, ISSN 0036-8075

Keller, H. U., Mottola, S., Davidsson, B., Schröder, S. E., Skorov, Y., Kührt, E., Groussin, O., Pajola, M., Hviid, S. F., Preusker, F., Scholten, F., A´Hearn, M. F., Sierks, H., Barbieri, C., Lamy, P., Rodrigo, R., Koschny, D., Rickman, H., Barucci, M. A., Bertaux, J.-L., Bertini, I., Cremonese, G., Da Deppo, V., Debei, S., De Cecco, M., Fornasier, S., Fulle, M., Gutiérrez, P. J., Ip, W.-H., Jorda, L., Knollenberg, J., Kramm, J. R., Küppers, M., Lara, L. M., Lazzarin, M., Lopez Moreno, J. J., Marzari, F., Michalik, H., Naletto, G., Sabau, L., Thomas, N., Vincent, J.-B., Wenzel, K.-P., Agarwal, J., Güttler, C., Oklay, N., Tubiana, C., 2015, Insolation, erosion, and morphology of comet 67P/Churyumov-Gerasimenko, Astronomy & Astrophysics, 583, A34 1 – A34 16

Kissel, J., Altwegg, K., Clark, B. C., Colangeli, L., Cottin, H., Czempiel, S., Eibl, J., Engrand, C., Fehringer, H. M., Feuerbacher, B., Fomenkova, M., Glasmachers, A., Greenberg, J. M., Grün, E., Haerendel, G., Henkel, H., Hilchenbach, M., von Hoerner, H., Höfner, H., Hornung, K., Jessberger, E. K., Koch, A., Krüger, H., Langevin, Y., Parigger, P., Raulin, F., Rüdenauer, F., Rynö, J., Schmid, E. R., Schulz, R., Silén, J., Steiger, W., Stephan, T., Thirkell, L., Thomas, R., Torkar, K., Utterback, N. G., Varmuza, K., Wanczek, K. P., Werther, W., Zscheeg, H., 2007, Cosima – high resolution time-of-flight secondary ion mass spectrometer for the analysis of cometary dust particles onboard rosetta, Space Science Reviews, 128, 823 – 867, ISSN 1572-9672

Klingelhöfer, G., Brückner, J., D'Uston, C., Gellert, R., Rieder, R., 2007, The Rosetta Alpha Particle X-Ray Spectrometer (APXS), Space Science Reviews, 128, 383 – 396, ISSN 1572-9672

Knapmeyer, M., Fischer, H.-H., Knollenberg, J., Seidensticker, K., Thiel, K., Arnold, W., Faber, C., Möhlmann, D., 2018, Structure and elastic parameters of the near surface of abydos site on comet 67P/Churyumov-Gerasimenko, as obtained by sesame/casse listening to the mupus insertion phase, Icarus, 310, 165 – 193, ISSN 0019-1035

Koenders, C., Glassmeier, K.-H., Richter, I., Motschmann, U., Rubin, M., 2013, Revisiting cometary bow shock positions, Planetary and Space Science, 87, 85 – 95, ISSN 0032-0633

Koenders, C., Goetz, C., Richter, I., Motschmann, U., Glassmeier, K.-H., 2016, Magnetic field pile-up and draping at intermediately active comets: results from comet 67P/Churyumov-Gerasimenko at 2.0 au, Monthly Notices of the Royal Astronomical Society, 462, 235 – 241

Kofman, W., Herique, A., Goutail, J.-P., Hagfors, T., Williams, I. P., Nielsen, E., Barriot, J.-P., Barbin, Y., Elachi, C., Edenhofer, P., Levasseur-Regourd, A.-C., Plettemeier, D., Picardi, G., Seu, R., Svedhem, V., 2007, The comet nucleus sounding experiment by radiowave transmission (consert): A short description of the instrument and of the commissioning stages, Space Science Reviews, 128, 413 – 432, ISSN 1572-9672

Kofman, W., Herique, A., Barbin, Y., Barriot, J.-P., Ciarletti, V., Clifford, S., Edenhofer, P., Elachi, C., Eyraud, C., Goutail, J.-P., Heggy, E., Jorda, L., Lasue, J., Levasseur-Regourd, A.-C., Nielsen, E., Pasquero, P., Preusker, F., Puget, P., Plettemeier, D., Rogez, Y., Sierks, H., Statz, C., Svedhem, H., Williams, I., Zine, S., Van Zyl, J., 2015, Properties of the 67P/Churyumov-Gerasimenko interior revealed by consert radar, Science, 349, ISSN 0036-8075

Krolikowska, M., 2003, 67P/Churyumov-Gerasimenko - potential target for the rosetta mission, Acta Astronautica, 53, 195 – 209

La Forgia, F., Lazzarin, M., Bodewits, D., A'Hearn, M. F., Bertini, I., Penasa, L., Naletto, G., Cremonese, G., Massironi, M., Ferri, F., Frattin, E., Lucchetti, A., Ferrari, S., Barbieri, C., 2017, Diurnal and seasonal variations of gas emissions in the inner coma of comet 67P/Churyumov-Gerasimenko observed with OSIRIS/Rosetta, European Planetary Science Congress, 11, 431

Lai, I.-L., Ip, W.-H., Lee, J.-C., Lin, Z.-Y., Vincent, J.-B., Oklay, N., Sierks, H., Barbieri, C., Lamy, P., Rodrigo, R., Koschny, D., Rickman, H., Keller, H. U., Agarwal, J., Barucci, M. A., Bertaux, J.-L., Bertini, I., Boudreault, S., Cremonese, G., Deppo, V. D., B., 2018, Seasonal variations in source regions of the dust jets on comet 67P/Churyumov-Gerasimenko, Astronomy & Astrophysics, in press

Lamy, P. L., Toth, I., Jorda, L., Weaver, H. A., A'Hearn, M., 1998, The nucleus and inner coma of Comet 46P/Wirtanen, Astronomy and Astrophysics, 335, 25 – 29

Lamy, P. L., Toth, I., Weaver, H. A., Jorda, L., Kaasalainen, M., Gutiérrez, P. J., 2006, Hubble space telescope observations of the nucleus and inner coma of comet 67P/Churyumov-Gerasimenko, Astronomy & Astrophysics, 458, 669 – 678

Lamy, P. L., Toth, I., Davidsson, B. J. R., Groussin, O., Gutiérrez, P., Jorda, L., Kaasalainen, M., Lowry, S. C., 2007, A portrait of the nucleus of comet 67P/Churyumov-Gerasimenko, Space Science Reviews, 128, 23 – 66, ISSN 1572-9672

Lancaster-Brown, P., 1985, Halley and His Comet, Blandford Press, ISBN 0713714476

Li, L.-Y., Wu, C.-Y., Thornton, C., 2001, A theoretical model for the contact of elasto-plastic bodies, Proceedings of the Institution of Mechanical Engineers, Part C: Journal of Mechanical Engineering Science, 216, 421 – 431

Lorek, S., Gundlach, B., Lacerda, P., Blum, J., 2016, Comet formation in collapsing pebble clouds - what cometary bulk density implies for the cloud mass and dust-to-ice ratio, Astronomy & Astrophysics, 587, A128 1 – A128 14

Mazelle, C., Neubauer, F. M., 1993, Discrete wave packets at the proton cyclotron frequency at comet p/halley, Geophysical Research Letters, 20, 153 – 156

Meier, P., Glassmeier, K.-H., Motschmann, U., 2016, Modified ion-weibel instability as a possible source of wave activity at comet 67P/Churyumov-Gerasimenko, Annales Geophysicae, 34, 691 – 707

Morbidelli, A., Rickman, H., 2015, Comets as collisional fragments of a primordial planetesimal disk, Astronomy & Astrophysics, 583, A43 1 – A43 9

Motschmann, U., Glassmeier, K.-H., 1993, Nongyrotropic distribution of pickup ions at comet P/Grigg-Skjellerup: A possible source of wave activity, Journal of Geophysical Research: Space Physics, 98, 20 977 – 20 983

Mottola, S., Arnold, G., Grothues, H.-G., Jaumann, R., Michaelis, H., Neukum, G., Bibring, J.-P., 2007, The rolis experiment on the rosetta lander, Space Science Reviews, 128, 241 – 255, ISSN 1572-9672

Mottola, S., Lowry, S., Snodgrass, C., Lamy, P. L., Toth, I., Rozek, A., Sierks, H., A´Hearn, M. F., Angrilli, F., Barbieri, C., Barucci, M. A., Bertaux, J.-L., Cremonese, G., Da Deppo, V., Davidsson, B., De Cecco, M., Debei, S., Fornasier, S., Fulle, M., Groussin, O., Gutiérrez, P., Hviid, S. F., Ip, W., Jorda, L., Keller, H. U., Knollenberg, J., Koschny, D., Kramm, R., Kührt, E., Küppers, M., Lara, L., Lazzarin, M., Lopez Moreno, J. J., Marzari, F., Michalik, H., Naletto, G., Rickman, H., Rodrigo, R., Sabau, L., Thomas, N., Wenzel, K.-P., Agarwal, J., Bertini, I., Ferri, F., Güttler, C., Magrin, S., Oklay, N., Tubiana, C., Vincent, J.-B., 2014, The rotation state of 67P/Churyumov-Gerasimenko from approach observations with the osiris cameras on rosetta, Astronomy & Astrophysics, 569, L2 1 – L2 5

Mottola, S., Arnold, G., Grothues, H.-G., Jaumann, R., Michaelis, H., Neukum, G., Bibring, J.-P., Schröder, S. E., Hamm, M., Otto, K. A., Pelivan, I., Proffe, G., Scholten, F., Tirsch, D., Kreslavsky, M., Remetean, E., Souvannavong, F., Dolives, B., 2015, The structure of the regolith on 67P/Churyumov-Gerasimenko from ROLIS descent imaging, Science, 349

Newton, I., 1687, Philosophiae naturalis principia mathematica, J. Societatis Regiae ac Typis J. Streater

Nilsson, H., Lundin, R., Lundin, K., Barabash, S., Borg, H., Norberg, O., Fedorov, A., Sauvaud, J.-A., Koskinen, H., Kallio, E., Riihelä, P., Burch, J. L., 2007, Rpc-ica: The ion composition analyzer of the rosetta plasma consortium, Space Science Reviews, 128, 671 – 695, ISSN 1572-9672

Nilsson, H., Stenberg Wieser, G., Behar, E., Wedlund, C. S., Gunell, H., Yamauchi, M., Lundin, R., Barabash, S., Wieser, M., Carr, C., Cupido, E., Burch, J. L., Fedorov, A., Sauvaud, J.-A., Koskinen, H., Kallio, E., Lebreton, J.-P., Eriksson, A., Edberg, N., Goldstein, R., Henri, P., Koenders, C., Mokashi, P., Nemeth, Z., Richter, I., Szego, K., Volwerk, M., Vallat, C., Rubin, M., 2015, Birth of a comet magnetosphere: A spring of water ions, Science, 347, ISSN 0036-8075

Nübold, H., Glassmeier, K.-H., 1999, Coagulation and accretion of magnetized dust: A source of remanent cometary magnetism?, Advances in Space Research, 24, 1163 – 1166, ISSN 0273-1177

Nübold, H., Glassmeier, K.-H., 2000, Accretional remanence of magnetized dust in the solar nebula, Icarus, 144, 149 – 159, ISSN 0019-1035

Ogliore, R., Butterworth, A., Fakra, S., Gainsforth, Z., Marcus, M., Westphal, A., 2010, Comparison of the oxidation state of fe in comet 81P/Wild 2 and chondritic-porous interplanetary dust particles, Earth and Planetary Science Letters, 296, 278 – 286, ISSN 0012-821X

Pätzold, M., Häusler, B., Aksnes, K., Anderson, J. D., Asmar, S. W., Barriot, J.-P., Bird, M. K., Boehnhardt, H., Eidel, W., Grün, E., Ip, W. H., Marouf, E., Morley, T., Neubauer, F. M., Rickman, H., Thomas, N., Tsurutani, B. T., Wallis, M. K., Wickramasinghe, N. C., Mysen, E., Olson, O., Remus, S., Tellmann, S., Andert, T., Carone, L., Fels, M., Stanzel, C., Audenrieth-Kersten, I., Gahr, A., Müller, A.-L., Stupar, D., Walter, C., 2007, Rosetta radio science investigations (rsi), Space Science Reviews, 128, 599 – 627, ISSN 1572-9672

Pätzold, M., Andert, T., Hahn, M., Asmar, S. W., Barriot, J. P., Bird, M. K., Häusler, B., Peter, K., Tellmann, S., Grün, E., Weissman, P. R., Sierks, H., Jorda, L., Gaskell, R., Preusker, F., Scholten, F., 2016, A homogeneous nucleus for comet 67P/Churyumov-Gerasimenko from its gravity field, Nature, 530, 63 – 65, ISSN 14764687

Poulet, F., Lucchetti, A., Bibring, J.-P., Carter, J., Gondet, B., Jorda, L., Langevin, Y., Pilorget, C., Capanna, C., Cremonese, G., 2016, Origin of the local structures at the philae landing site and possible implications on the formation and evolution of 67P/Churyumov-Gerasimenko, Monthly Notices of the Royal Astronomical Society, 462, 23 – 32

Preusker, F., Scholten, F., Matz, K.-D., Roatsch, T., Willner, K., Hviid, S. F., Knollenberg, J., Jorda, L., Gutiérrez, P. J., Kührt, E., Mottola, S., A´Hearn, M. F., Thomas, N., Sierks, H., Barbieri, C., Lamy, P., Rodrigo, R., Koschny, D., Rickman, H., Keller, H. U., Agarwal, J., Barucci, M. A., Bertaux, J.-L., Bertini, I., Cremonese, G., Da Deppo, V., Davidsson, B., Debei, S., De Cecco, M., Fornasier, S., Fulle, M., Groussin, O., Güttler, C., Ip, W.-H., Kramm, J. R., Küppers, M., Lara, L. M., Lazzarin, M., Lopez Moreno, J. J., Marzari, F., Michalik, H., Naletto, G., Oklay, N., Tubiana, C., Vincent, J.-B., 2015, Shape model, reference system definition, and cartographic mapping standards for comet 67P/Churyumov-Gerasimenko - stereo-photogrammetric analysis of rosetta/osiris image data, Astronomy & Astrophysics, 583, A33 1 – A33 19

Preusker, F., Scholten, F., Matz, K.-D., Roatsch, T., Hviid, S. F., Mottola, S., Knollenberg, J., Kührt, E., Pajola, M., Oklay, N., Vincent, J.-B., Davidsson, B., A´Hearn, M. F., Agarwal, J., Barbieri, C., Barucci, M. A., Bertaux, J.-L., Bertini, I., Cremonese, G., Da Deppo, V., Debei, S., De Cecco, M., Fornasier, S., Fulle, M., Groussin, O., Gutiérrez, P. J., Güttler, C., Ip, W-H., Jorda, L., Keller, H. U., Koschny, D., Kramm, J. R., Küppers, M., Lamy, P., Lara, L. M., Lazzarin, M., Lopez Moreno, J. J., Marzari, F., Massironi, M., Naletto, G., Rickman, H., Rodrigo, R., Sierks, H., Thomas, N., Tubiana, C., 2017, The global meter-level shape model of comet 67P/Churyumov-Gerasimenko, Astronomy & Astrophysics, 607, L1 1 – L1 5

Remetean, E., Dolives, B., Souvannavong, F., Germa, T., Ginestet, J., Torres, A., Mousset, T., 2016, Philae locating and science support by robotic vision techniques, Acta Astronautica, 125, 161 – 173, ISSN 0094-5765

Richter, I., Koenders, C., Glassmeier, K., Tsurutani, B., Goldstein, R., 2011, Deep space 1 at comet 19p/borrelly: Magnetic field and plasma observations, Planetary and Space Science, 59, 691 – 698, ISSN 0032-0633

Richter, I., Auster, H.-U., Glassmeier, K.-H., Koenders, C., Carr, C., Motschmann, U., Müller, J., McKenna-Lawlor, S., 2012, Magnetic field measurements during the rosetta flyby at asteroid (21)lutetia, Planetary and Space Science, 66, 155 – 164, ISSN 0032-0633, rosetta Fly-by at Asteroid (21) Lutetia

Richter, I., Koenders, C., Auster, H.-U., Frühauff, D., Götz, C., Heinisch, P., Perschke, C., Motschmann, U., Stoll, B., Altwegg, K., Burch, J., Carr, C., Cupido, E., Eriksson, A., Henri, P., Goldstein, R., Lebreton, J.-P., Mokashi, P., Nemeth, Z., Nilsson, H., Rubin, M., Szegö, K., Tsurutani, B. T., Vallat, C., Volwerk, M., Glassmeier, K.-H., 2015, Observation of a new type of low-frequency waves at comet 67P/Churyumov-Gerasimenko, Annales Geophysicae, 33, 1031 – 1036

Richter, I., Auster, H.-U., Berghofer, G., Carr, C., Cupido, E., Fornaçon, K.-H., Goetz, C., Heinisch, P., Koenders, C., Stoll, B., Tsurutani, B. T., Vallat, C., Volwerk, M., Glassmeier, K.-H., 2016, Two-point observations of low-frequency waves at 67P/Churyumov-Gerasimenko during the descent of philae: comparison of rpcmag and romap, Annales Geophysicae, 34, 609 – 622

Riedler, W., Torkar, K., Jeszenszky, H., Romstedt, J., Alleyne, H. S. C., Arends, H., Barth, W., Biezen, J. V. D., Butler, B., Ehrenfreund, P., Fehringer, M., Fremuth, G., Gavira, J., Havnes, O., Jessberger, E. K., Kassing, R., Klöck, W., Koeberl, C., Levasseur-Regourd, A. C., Maurette, M., Rüdenauer, F., Schmidt, R., Stangl, G., Steller, M., Weber, I., 2007, MIDAS - The Micro-Imaging Dust Analysis System for the Rosetta Mission, Space Science Reviews, 128, 869 – 904, ISSN 1572-9672

Rogez, Y., Puget, P., Zine, S., Hérique, A., Kofman, W., Altobelli, N., Ashman, M., Barthelemy, M., Biele, J., Blazquez, A., Casas, C. M., Sitjà, M. C., Delmas, C., Fantinati, C., Fronton, J.-F., Geiger, B., Geurts, K., Grieger, B., Hahnel, R., Hoofs, R., Hubault, A., Jurado, E., Küppers, M., Maibaum, M., Moussi-Souffys, A., Muñoz, P., O'Rourke, L., Pätz, B., Plettemeier, D., Ulamec, S., Vallat, C., 2016, The {CONSERT} operations planning process for the Rosetta mission, Acta Astronautica, 125, 212 – 233, ISSN 0094-5765

Roll, R., Witte, L., 2016, Rosetta lander philae: Touch-down reconstruction, Planetary and Space Science, 125, 12 – 19, ISSN 0032-0633

Roll, R., Witte, L., Arnold, W., 2016, Rosetta lander philae – soil strength analysis, Icarus, 280, 359 – 365, ISSN 0019-1035, microMars to MegaMars

Sauer, K., Dubinin, E., Baumgärtel, K., Tarasov, V., 1998, Low-frequency electromagnetic waves and instabilities within the martian bi-ion plasma, Earth, Planets and Space, 50, 269 – 278, ISSN 1880-5981

Schräpler, R., Blum, J., von Borstel, I., Güttler, C., 2015, The stratification of regolith on celestial objects, Icarus, 257, 33 – 46, ISSN 0019-1035

Schröder, S., Mottola, S., Arnold, G., Grothues, H.-G., Jaumann, R., Keller, H., Michaelis, H., Bibring, J.-P., Pelivan, I., Koncz, A., Otto, K., Remetean, E., Souvannavong, F., Dolives, B., 2017, Close-up images of the final philae landing site on comet 67P/Churyumov-Gerasimenko acquired by the rolis camera, Icarus, 285, 263 – 274, ISSN 0019-1035

Schulz, R., 2010, The Rosetta mission and its fly-by at asteroid 2867 Steins, Planetary and Space Science, 58, 1057, ISSN 0032-0633, special Issue: Rosetta Fly-by at Asteroid (2867) Steins

Schulz, R., Sierks, H., Küppers, M., Accomazzo, A., 2012, Rosetta fly-by at asteroid (21) lutetia: An overview, Planetary and Space Science, 66, 2 – 8, ISSN 0032-0633, rosetta Fly-by at Asteroid (21) Lutetia

Seidensticker, K. J., Möhlmann, D., Apathy, I., Schmidt, W., Thiel, K., Arnold, W., Fischer, H.-H., Kretschmer, M., Madlener, D., Péter, A., Trautner, R., Schieke, S., 2007, Sesame – An Experiment of the Rosetta Lander Philae: Objectives and General Design, Space Science Reviews, 128, 301 – 337, ISSN 1572-9672

Shumway, R. H., Stoffer, D. S., 2011, Spectral Analysis and Filtering, pp. 173 – 265, Springer New York, New York, NY, ISBN 978-1-4419-7865-3

Sierks, H., 2016, PHILAE FOUND!, ESA ROSETTA Blog, http://blogs.esa.int/rosetta/2016/09/05/philae-found/;

Smith, E. J., Tsurutani, B. T., 1976, magnetosheath lion roars, Journal of Geophysical Research, 81, 2261 – 2266, ISSN 2156-2202

Smith, E. J., Tsurutani, B. T., Slavin, J. A., Jones, D. E., Siscoe, G. L., Mendis, D. A., 1986, International cometary explorer encounter with giacobini-zinner: Magnetic field observations, Science, 232, 382 – 385, ISSN 0036-8075

Snodgrass, C., Tubiana, C., Bramich, D. M., Meech, K., Boehnhardt, H., Barrera, L., 2013, Beginning of activity in 67P/Churyumov-Gerasimenko and predictions for 2014-2015, Astronomy & Astrophysics, 557, A33 1 – A33 15

Sonnerup, B. U., Cahill, L. J., 1967, Magnetopause structure and attitude from explorer 12 observations, Journal of Geophysical Research, 72, 171 – 183, ISSN 2156-2202

Sonnerup, B. U. Ö., Scheible, M., 1998, Minimum and Maximum Variance Analysis, ISSI Scientific Reports Series, 1, 185 – 220

Spohn, T., Seiferlin, K., Hagermann, A., Knollenberg, J., Ball, A. J., Banaszkiewicz, M., Benkhoff, J., Gadomski, S., Gregorczyk, W., Grygorczuk, J., Hlond, M., Kargl, G., Kührt, E., Kömle, N., Krasowski, J., Marczewski, W., Zarnecki, J. C., 2007, Mupus – A Thermal and Mechanical Properties Probe for the Rosetta Lander Philae, Space Science Reviews, 128, 339 – 362, ISSN 1572-9672

Spohn, T., Knollenberg, J., Ball, A. J., Banaszkiewicz, M., Benkhoff, J., Grott, M., Gry-gorczuk, J., Hüttig, C., Hagermann, A., Kargl, G., Kaufmann, E., Kömle, N., Kührt, E., Kossacki, K. J., Marczewski, W., Pelivan, I., Schrödter, R., Seiferlin, K., 2015, Thermal and mechanical properties of the near-surface layers of comet 67P/Churyumov-Gerasimenko, Science, 349, ISSN 0036-8075

Stenberg Wieser, G., Odelstad, E., Wieser, M., Nilsson, H., Goetz, C., Karlsson, T., André, M., Kalla, L., Eriksson, A. I., Nicolaou, G., Wedlund, C. S., Richter, I., Gunell, H., 2017, Investigating short-time-scale variations in cometary ions around comet 67p, Monthly Notices of the Royal Astronomical Society, 469, 522 – 534

Stern, S. A., Slater, D. C., Scherrer, J., Stone, J., Versteeg, M., A'hearn, M. F., Bertaux, J. L., Feldman, P. D., Festou, M. C., Parker, J. W., Siegmund, O. H. W., 2007, Alice: The rosetta ultraviolet imaging spectrograph, Space Science Reviews, 128, 507 – 527, ISSN 1572-9672

Szegö, K., Glassmeier, K.-H., Bingham, R., Bogdanov, A., Fischer, C., Haerendel, G., Brinca, A., Cravens, T., Dubinin, E., Sauer, K., Fisk, L., Gombosi, T., Schwadron, N., Isenberg, P., Lee, M., Mazelle, C., Möbius, E., Motschmann, U., Shapiro, V. D., Tsurutani, B., Zank, G., 2000, Physics of mass loaded plasmas, Space Science Reviews, 94, 429 – 671, ISSN 1572-9672

Thiel, M., Stöcker, J., Rohe, C., Kömle, N. I., Kargl, G., Hillenmaier, O., Lell, P., 2003, The ROSETTA Lander anchoring system, in: 10th European Space Mechanisms and Tribology Symposium, (Ed.) R. A. Harris, vol. 524 of ESA Special Publication, pp. 239 – 246

Thomas, N., Davidsson, B., El-Maarry, M. R., Fornasier, S., Giacomini, L., Gracia-Berná, A. G., Hviid, S. F., Ip, W.-H., Jorda, L., Keller, H. U., Knollenberg, J., Kührt, E., La Forgia, F., Lai, I. L., Liao, Y., Marschall, R., Massironi, M., Mottola, S., Pajola, M., Poch, O., Pommerol, A., Preusker, F., Scholten, F., Su, C. C., Wu, J. S., Vincent, J.-B., Sierks, H., Barbieri, C., Lamy, P. L., Rodrigo, R., Koschny, D., Rickman, H., A´Hearn, M. F., Barucci, M. A., Bertaux, J.-L., Bertini, I., Cremonese, G., Da Deppo, V., Debei, S., de Cecco, M., Fulle, M., Groussin, O., Gutierrez, P. J., Kramm, J.-R., Küppers, M., Lara, L. M., Lazzarin, M., Lopez Moreno, J. J., Marzari, F., Michalik, H., Naletto, G., Agarwal, J., Güttler, C., Oklay, N., Tubiana, C., 2015, Redistribution of particles across the nucleus of comet 67P/Churyumov-Gerasimenko, Astronomy & Astrophysics, 583, A17 1 – A17 18

Trotignon, J. G., Michau, J. L., Lagoutte, D., Chabassière, M., Chalumeau, G., Colin, F., Décréau, P. M. E., Geiswiller, J., Gille, P., Grard, R., Hachemi, T., Hamelin, M., Eriksson, A., Laakso, H., Lebreton, J. P., Mazelle, C., Randriamboarison, O., Schmidt, W., Smit, A., Telljohann, U., Zamora, P., 2007, Rpc-mip: the mutual impedance probe of the rosetta plasma consortium, Space Science Reviews, 128, 713 – 728, ISSN 1572-9672

Tsurutani, B. T., Smith, E. J., 1986, Strong hydromagnetic turbulence associated with comet giacobini-zinner, Geophysical Research Letters, 13, 259 – 262

Tsurutani, B. T., Thorne, R. M., Smith, E. J., Gosling, J. T., Matsumoto, H., 1987, Steepened magnetosonic waves at Comet Giacobini-Zinner, Journal of Geophysical Research, 92, 11 074 – 11 082

Ulamec, S., Taylor, M. G., 2016, Editorial of the special issue – "rosetta and philae at comet 67P/Churyumov-Gerasimenko", Acta Astronautica, 125, 1 – 2, ISSN 0094-5765

Ulamec, S., Espinasse, S., Feuerbacher, B., Hilchenbach, M., Moura, D., Rosenbauer, H., Scheuerle, H., Willnecker, R., 2006, Rosetta Lander—Philae: Implications of an alternative mission, Acta Astronautica, 58, 435 – 441, ISSN 0094-5765

Ulamec, S., Biele, J., Fantinati, C., Fronton, J.-F., Gaudon, P., Geurts, K., Krause, C., Küchemann, O., Maibaum, M., Pätz, B., Roll, R., Willnecker, R., 2012, Rosetta lander—after seven years of cruise, prepared for hibernation, Acta Astronautica, 81, 151 – 159, ISSN 0094-5765

Ulamec, S., Biele, J., Blazquez, A., Cozzoni, B., Delmas, C., Fantinati, C., Gaudon, P., Geurts, K., Jurado, E., Küchemann, O., Lommatsch, V., Maibaum, M., Sierks, H., Witte, L., 2015, Rosetta Lander – Philae: Landing preparations, Acta Astronautica, 107, 79 – 86, ISSN 0094-5765

Ulamec, S., Fantinati, C., Maibaum, M., Geurts, K., Biele, J., Jansen, S., Küchemann, O., Cozzoni, B., Finke, F., Lommatsch, V., Moussi-Soffys, A., Delmas, C., O'Rourke, L., 2016, Rosetta Lander – Landing and operations on comet 67P/Churyumov-Gerasimenko, Acta Astronautica, 125, 80 – 91, ISSN 0094-5765

Ulamec, S., O'Rourke, L., Biele, J., Grieger, B., Andrés, R., Lodiot, S., Muñoz, P., Charpentier, A., Mottola, S., Knollenberg, J., Knapmeyer, M., Kührt, E., Scholten, F., Geurts, K., Maibaum, M., Fantinati, C., Küchemann, O., Lommatsch, V., Delmas, C., Jurado, E., Garmier, R., Martin, T., 2017, Rosetta Lander - Philae: Operations on comet 67P/Churyumov-Gerasimenko, analysis of wake-up activities and final state, Acta Astronautica, 137, 38 – 43, ISSN 0094-5765

Vazquez-Garcia, J., Zender, J., Barthelemy, M., Semenov, B., 2015, ROSETTA OR-BITER/LANDER SPICE KERNELS V1.0, RO/RL-E/M/A/C-SPICE-6-V1.0, ftp: //ssols01.esac.esa.int/pub/data/SPICE/ROSETTA/kernels/;

Verkhoglyadova, O. P., Tsurutani, B. T., Lakhina, G. S., 2013, Theoretical analysis of Poynting flux and polarization for ELF-VLF electromagnetic waves in the Earth's magnetosphere, Journal of Geophysical Research: Space Physics, 118, 7695 – 7702, ISSN 2169-9402

Volwerk, M., Glassmeier, K.-H., Delva, M., Schmid, D., Koenders, C., Richter, I., Szegö, K., 2014, A comparison between VEGA 1, 2 and Giotto flybys of comet 1P/Halley: implications for Rosetta, Annales Geophysicae, 32, 1441 – 1453

Volwerk, M., Goetz, C., Richter, I., Delva, M., Ostaszewski, K., Schwingenschuh, K., Glassmeier, K.-H., 2018, A tail like no other - The RPC-MAG view of Rosetta's tail excursion at comet 67P/Churyumov-Gerasimenko, Astronomy & Astrophysics, 614, A10 1 – A10 10

Wahlberg Jansson, K., Johansen, A., 2014, Formation of pebble-pile planetesimals, Astronomy & Astrophysics, 570, A47 1 – A47 10

Wahlberg Jansson, K., Johansen, A., Bukhari, M., Jürgen Blum, J., 2017, The role of pebble fragmentation in planetesimal formation. ii. numerical simulations, The Astrophysical Journal, 835, 109 1 – 109 11

Wang, H., Weiss, B. P., Bai, X.-N., Downey, B. G., Wang, J., Wang, J., Suavet, C., Fu, R. R., Zucolotto, M. E., 2017, Lifetime of the solar nebula constrained by meteorite paleomagnetism, Science, 355, 623 – 627, ISSN 0036-8075

Welch, P., 1967, The use of fast fourier transform for the estimation of power spectra: A method based on time averaging over short, modified periodograms, IEEE Transactions on Audio and Electroacoustics, 15, 70 – 73, ISSN 0018-9278

Witte, L., Schroeder, S., Kempe, H., van Zoest, T., Roll, R., Ulamec, S., Biele, J., Block, J., 2014, Experimental investigations of the comet lander philae touchdown dynamics, Journal of Spacecraft and Rockets, 51, 1885 – 1894, ISSN 0022-4650

Witte, L., Roll, R., Biele, J., Ulamec, S., Jurado, E., 2016, Rosetta lander Philae – Landing performance and touchdown safety assessment, Acta Astronautica, 125, 149 – 160, ISSN 0094-5765

Wright, I. P., Barber, S. J., Morgan, G. H., Morse, A. D., Sheridan, S., Andrews, D. J., Maynard, J., Yau, D., Evans, S. T., Leese, M. R., Zarnecki, J. C., Kent, B. J., Waltham, N. R., Whalley, M. S., Heys, S., Drummond, D. L., Edeson, R. L., Sawyer, E. C., Turner, R. F., Pillinger, C. T., 2007, Ptolemy – an instrument to measure stable isotopic ratios of key volatiles on a cometary nucleus, Space Science Reviews, 128, 363 – 381, ISSN 1572-9672

Wu, C. S., Davidson, R. C., 1972, Electromagnetic instabilities produced by neutral-particle ionization in interplanetary space, Journal of Geophysical Research, 77, 5399 – 5406

Yigit, A. S., Christoforou, A. P., Majeed, M. A., 2011, A nonlinear visco-elastoplastic impact model and the coefficient of restitution, Nonlinear Dynamics, 66, 509 – 521, ISSN 1573-269X

Danksagung

An dieser Stelle möchte ich mich bei allen bedanken, die mir diese Arbeit ermöglicht haben. Besonders möchte ich Prof. Dr. Karl-Heinz Glaßmeier für die Begleitung dieser Arbeit als Mentor und die kontinuierliche Förderung seit der Bachelor Arbeit danken. Über die rein wissenschaftliche Unterstützung hinaus hat er es mir immer wieder ermöglicht, im Rahmen von zahlreichen In- und Auslandsdienstreisen wissenschaftliche Ergebnisse zu präsentieren und sich mit Kollegen auszutauschen.

Mein Dank gebührt auch Dr. Hans-Ulrich Auster, der mir die Möglichkeit gegeben hat im ROMAP Team mitzuarbeiten und mir immer stets hilfreich zur Seite stand. Ohne ihn wäre auch nie die Idee entstanden, dass Magnetfeldmessungen in irgendeiner Form dabei helfen könnten, etwas über die mechanischen Eigenschaften eines Kometen zu erfahren. Ohne die zahlreichen intensiven fachlichen Diskussionen wäre es in dieser Form weder zu den vorangegangenen Veröffentlichungen noch zu dieser Arbeit gekommen.

Dr. Ingo Richter war nicht nur eine große Hilfe bei der Auswertung der RPC-MAG Messdaten im Rahmen dieser Arbeit, sondern auch bei zahlreichen gemeinsamen Veröffentlichungen. Er war darüber hinaus ein ständiger Ansprechpartner für alle Fragen rund um die Physik, Elektrotechnik und Luft- und Raumfahrt und half mir stets mit Rat und Tat.

Auch Katharina Ostaszewski möchte ich sowohl für die intensiven wissenschaftlichen Diskussionen als auch für das Korrekturlesen dieser Arbeit danken. Besonders der intensive Austausch zur Beschreibung und Simulation von Wellenphänomenen in Plasmen haben meine Sicht auf die Plasmaphysik und die Numerik nachhaltig verändert.

Bei Dr. Bastian Gundlach möchte ich mich für die vielen Informationen und Laborergebnisse zur Entstehung und den mechanischen Eigenschaften von Kometen bedanken. Er hat mir die relativ kurzfristige Einarbeitung in dieses Fachgebiet im Rahmen dieser Arbeit erheblich erleichtert.

Dem ganzen Rosetta Team, insbesondere den vielen Kollegen vom RPC, DLR-LCC, SONC, SGS und RMOC möchte ich für den Erfahrungsaustausch, der Bereitstellung zahlreicher technischer Daten und Informationen und die jahrelange Unterstützung danken. Ohne dieses Team wäre weder die Rosetta Mission noch diese Arbeit möglich gewesen.

Weiterhin möchte ich meinen Eltern danken, ohne deren kontinuierliche Unterstützung hätte es diese Arbeit nie gegeben.

www.ingramcontent.com/pod-product-compliance
Lightning Source LLC
Chambersburg PA
CBHW052109230326
41599CB00054B/5271